TRAILBLAZING

The Quest for Energy Self-Reliance

TRAILBLAZING
The Quest for Energy Self-Reliance

Geronimo Z. Velasco

Edited by Maricris R. Valte

Anvil
Manila

Copyright © by Geronimo Z. Velasco, 2006

All rights reserved.
No copies can be made in part or in whole without prior written permission from the author and the publisher.

Published and exclusively distributed by
ANVIL PUBLISHING, INC.
8007-B Pioneer St., Bgy. Kapitolyo
Pasig City 1603 Philippines
Sales and Marketing: 6373621, 6375141, 7471624;
marketing@anvilpublishing.com
Fax: 6376084
url: www.anvilpublishing.com

First printing, 2006

The National Library of the Philippines CIP Data

Recommended entry:

Velasco, Geronimo Z.
    Trailblazing: The Quest for Energy Self-Reliance / Geronimo Z. Velasco. – Pasig City: Anvil Pub., c2006.
      1v

    1. Velasco, Geronimo Z. 2. Power resources-Personal narratives. 3. Philippine National Oil Company-Politics and government-Personal narratives. 4. Public administration-Philippines-Personal narratives. I. Title

HD9502 354.4092 2006 P061000004
ISBN 971-27-1771-2

Editorial Consultant: Laura L. Samson
Copyeditor: Jocelyn de Jesus
Book Design: Veni L. Ilowa and Dezh David
Cover Design: Dezh David

Printed in the Philippines by COR ASIA, Inc.

To my wife, Erlinda
– for all my shortcomings

To my children
Banjo, Penny, Jerome, Marisa, and Miguel –
hoping that you will also be given an opportunity to help
solve some of our country's problems

Contents

Preface ... ix

Chapter 1
Confronting the Oil Crisis ... 1

Chapter 2
Working for Energy Self-Reliance ... 33

Chapter 3
The Tragedy of the Bataan Nuclear Power Plant ... 89

Chapter 4
Privatizing Power Generation: The Tale of NAPOCOR ... 131

Chapter 5
Assessing the Role of Foreign Oil Companies ... 161

Postscript ... 189

Preface

For someone who thinks memoirs are nothing but bravura talk and a lot of chest-beating, I am writing mine.

In December 1973, I reached what was to be the peak of my professional career when President Ferdinand Marcos appointed me as president and chief executive officer of the Philippine National Oil Company (PNOC). Three years later, I was named secretary of the newly created Department of Energy, which was tasked with the responsibility of developing the first ever "Total Energy Plan" for the country.

Looking back, I realized that driving in the fast lane had always been the pattern of my professional life, with career development, promotions and the climb up the ladder achieved in a fairly short time. My own professional career spanned nearly forty years. I worked myself to the top, but the ride was not that easy. While I started out armed with an excellent academic background, having graduated from the Mapua Institute of Technology as a member of the Mechanical Engineering Class of 1951 and topping the board examinations, my only real asset was myself.

Like all young men, I was driven and determined to reach greater heights, never afraid to delve into the unfamiliar and always on the lookout for challenges to conquer. But unlike all young men, I had the fortune of being at the right place at the right time. I must say I had fantastic timing, which was neither planned nor deliberate. But luck is an added bonus to my career. Getting to the top is one thing, but staying there requires something more than luck.

At thirty-four, I became president of Republic Glass Corporation—my first big professional break. Three years later, in 1965, I became concurrently vice president and, in 1967, president of Dole (Philippines), besides sitting in the board of directors of major corporations like General

Motors (Philippines), RCA (Philippines), Rizal Commercial Banking Corporation, and several others. I was also chairman of Castle and Cooke's subsidiaries in Thailand and Singapore. This earned me the reputation of being the country's highest-paid executive, whether Filipino or foreign.

At work, I have always endeavored to reach the highest standard of excellence and have always been enthusiastic to meet professional challenges.

The highlight of my career came when I was named president and CEO of PNOC, and subsequently secretary of energy. The shift from private to public sector occurred when the whole world was reeling from the first "oil shock" in the early Seventies. During this time, I had the responsibility of putting together the first comprehensive energy plan for the country. I tried to give total devotion and commitment to this task. For someone who took the job with only a vague familiarity with energy, I can say I managed to steer the development of our energy plan toward the goal of veering the country away from oil dependency and tapping alternative energy sources like geothermal, coal, and even the controversial nuclear power plant.

During my term from 1973 to 1985, it was a dream for many aspiring executives and engineers to work in PNOC. After all, ours was the only Filipino-owned corporation ever to become listed on *Fortune* magazine's Top 500 Companies outside the United States from 1978 to 1981. At a time when having a national oil company was considered an important policy instrument to aid governments in dealing with volatile oil markets, PNOC was held up by independent observers as worthy of emulation. The World Bank lauded our energy program as a model for Third World countries. When the international consulting firm Arthur D. Little made an evaluation study (funded by the Asian Development Bank) for the Petroleum Authority of Thailand in 1985, among its recommendations was for the latter to follow the organizational pattern and system of PNOC. Yet, despite having met the challenges of two oil crises, I remained just as zealous in making sure that PNOC and the Ministry of Energy kept to their mission of weaning the country's energy consumption away from oil and achieving energy self-reliance.

But my greatest task was cut short with the change in government in February 1986, just when we were about to see the fruition of our efforts.

Upon my designation as PNOC's president and CEO, one of the first persons to congratulate me was my good friend Leonides Virata. To this day, I cannot forget what he said: "Ronnie, you have a good opportunity to serve the country. But remember, just like in economics, government service has a point of diminishing returns. The key is for you to determine when to get out before you reach that point."

To me, that point was reached in June 1985, when the *San Jose (California) Mercury News* published an article that questioned my integrity, when it claimed that I had acquired a residential property in California under questionable circumstances, implying that I had acquired ill-gotten wealth. But what was more frustrating was that I could not vent my ire on a foreign publication. I wanted to sue *Mercury News* for libel, but my lawyer-friends advised against it. In other words, I had to swallow everything.

Besides, at that point I could already sense that an anti-Marcos sentiment was beginning to snowball in the American press. Under the circumstances, I thought that it was best to tender my resignation. The only way I could combat the smear campaign of President Marcos's foes was for me to leave government service, and the only means of defense I had was to challenge my accusers to ask all the people I had worked with as to whether I committed graft in order to buy that house in California.

My anger stemmed from the fact that the reporter from *Mercury News* even interviewed me for the article. He came to my house in Dasmariñas Village and I explained to him (in the presence of Boo Chanco, who worked with me at PNOC and the ministry) my personal circumstances and corporate career prior to entering government. But apparently, as it turned out from the article, the reporter had already formed his own conclusions and was out to perform a demolition job.

I tendered my resignation on July 1, 1985, with a heavy heart (see full text of my resignation letter on pages xx to xxiii). As I had written to President Marcos, although false reports and smear campaigns may be "facts of life" for those actively involved in politics, I was not a politician and had no ambition of becoming one. I could not stomach the indignity that malicious talks heaped on me and hence my family. Given the strengthening anti-Marcos opposition, I also felt that I would be a convenient target for the president's opponents, and this was a situation that I felt ill-equipped to handle. As a professional manager, I also believed that my effectiveness had already been compromised by *Mercury News*. I realized that my resignation could be taken to mean guilt but, as I wrote to the PNOC staff, *"After all is said and done, I am at peace with the thought that my best witnesses are the people I have worked with these past twelve years."*[1]

Unfortunately, President Marcos did not accept my resignation. He sent two of his closest aides as emissaries, with the message, *"Sabihin niyo kay Ronnie, para naman kaming walang pinagsamahan* (Tell Ronnie it's as if we hadn't been through good times and bad)." That broke my resolve to quit. It is true, the President and I had known each other for nearly twenty years, and all throughout my stint in government he gave me unflinching support. How could I now turn my back on him, at a time when he needed support the most? It was simply against Filipino culture to abandon the person who gave me a tremendous opportunity to serve the country. At the same time, I could not help but think that, maybe, President Marcos wanted me to let the political storm pass before leaving his Cabinet.

As if to show his continued confidence in me, after the snap elections in February 1986, the president reappointed me as energy minister. I remember then-Trade Minister Roberto "Bobby" Ongpin telling me that I was first on the list of Cabinet members to be retained. It was obvious that I could not back out anymore, so I pleaded with Bobby to convince the president to give me another job, since I had been dealing with energy for twelve years and I wanted to move on. We did not get to that point, of course, since events overtook us. The Edsa rebellion against President Marcos in February 1986 put an end to his presidency and thus to my career in government.

---

1. "Message from the Chairman," *Sandiwa ng PNOC*, Vol. IX No. 8, July 15, 1985.

It has been twenty years since then, so it is fair to ask, "Why talk about your experiences now?"

Immediately upon leaving office in 1986, my impulse was to write a book about our accomplishments at PNOC and the Ministry of Energy. The purpose was not so much to crow about them (although we had every good reason to) but to neutralize the efforts of those who took over PNOC to completely negate the work we had done in earnest for over a decade. I was also motivated by the need to defend myself from accusations that our ministry was the "most corrupt" and that I was unfaithful to the causes of proper governance. I could not accept these attacks on my personal integrity and the sincerity with which I governed PNOC and the ministry.

It was most disheartening, to say the least. Of course, at that particular time, I had to grit my teeth in silence, as the prevailing *zeitgeist* did not even allow me to defend myself without being accused of self-interestedness. No matter how I cried to the high heavens, no one would believe me. I also realized that the more I tried to explain, the more people became cynical toward me. I consulted my family and those close to me, and they all agreed that telling the real story would have to wait until such time when the passion of political partisanship had subsided sufficiently to allow for an unbiased assessment of the past.

In the meantime, I attended to the criminal and civil cases filed against me by the Presidential Commission on Good Government (PCGG) before the Ombudsman and the Sandiganbayan. I was acquitted in all the criminal cases. In subsequent appeals made before the Supreme Court of the Philippines, the latter upheld the initial verdict of the Sandiganbayan that I was innocent. As for the civil cases filed against my person, as of this writing there is still one case pending before the sala of the late Sandiganbayan Presiding Justice Francis Garchitorena, who was suspended by the Supreme Court for refusing to act on cases heard in his sala and subsequently sought early retirement. Among the cases that were deliberately left undecided was mine. In my desire to expedite the resolution of the case, I even gave advance testimony before the Sandiganbayan on June 19, 1995. Since then, no subsequent hearing has been held and, more than ten years after that testimony, the case remains unresolved.

Let me just relate two examples of how PCGG went overboard in trying to prove my guilt. First, they tried to sequester my house in Dasmariñas Village for supposedly being ill-gotten. What they did not know was that I built that house in 1968, but PCGG agents insisted that I bought it only in 1984. It was a good thing that then-PCGG Chairman Jovito Salonga had been to my residence a number of times before martial law, and so he corroborated my statement. The PCGG investigators also pressured the executives and staff of PNOC and the ministry, trying to extract "confessions" that would prove my guilt over the various corruption charges. I am proud to say that, despite the emotional and psychological harassment they had to undergo, my former colleagues stood their ground and refused to go along with the charade. They proved me right, when I said that "my best witnesses are the people I have worked with these past twelve years." I am also proud of the fact that not a single asset of mine was ever sequestered.

Despite my decision to maintain a public silence, all these years I kept as much material on PNOC's history as I could. Many important documents (some of them personal ones) remain in the PCGG's custody. Fortunately, immediately after leaving office and while staying in San Francisco, I managed to dictate to my former secretary my recollections and reflections that I thought may have to be revealed someday, either to defend my cases in court or to clarify the early history of PNOC. The transcripts provided the notes that enabled me to talk about my years in government despite the lack of official documents.

I am writing this book primarily as *my way of paying tribute to the men and women of PNOC and the Ministry of Energy, who worked just as hard and with genuine passion and commitment to realize our vision for the country. They showed that government service could be on a par with the best in the private sector. Without them, I would not have been able to steer PNOC toward attaining the many accomplishments that remain unsurpassed in government annals to this day.* I think it is fair to say that all of us gave the best years of our lives to PNOC.

I had been an executive of major Philippine corporations, and my experience and exposure to these companies prepared me for the challenges that awaited me upon assuming responsibility for PNOC and the Ministry of Energy. My professional life had been so blessed with good fortune that my brush with a near-ending in 1986 took a toll professionally and personally. The stigma of being associated with President Marcos erased all the hard work I had put in as a public servant. Nothing pained me more than to be accused of being a traitor to my country. It was a blow that struck so hard that, were it not for my sense of humor and the support of my family, I would not have been able to hurdle that difficult experience.

My wife, Erlinda, was in San Francisco at the time of the Edsa rebellion, which was a good thing. Fortunately, too, my children were all grown-up by then. My eldest son Banjo, and daughters Marisa and Penny, were already married; Jerome was about to enter the MBA program at Wharton; and my youngest, Miguel, was about to finish his baccalaureate at Pepperdine. I already had three grandchildren, who were of preschool age. Although the sudden turn of events affected all of us in one way or another, my forced temporary residence in the United States gave me an opportunity to spend more quality time with my family.

My transition from being one of the most prominent to being one of the most maligned men in the Marcos administration was difficult. For someone who had been used to staying at the top and being in command, my hasty departure from office under ignominious conditions would have broken my spirit, as it would anyone used to wielding power and enjoying the trappings of power. When I returned to Manila in October 1986, I could sense that many people whom I used to socialize or play golf with had changed. The warmth was no longer there. In social gatherings, people would stare whenever I walked into the room, as if I had a contagious disease. It was during that time when I came to know who my real friends were.

I also felt the change in lifestyle. I cannot deny the glamor and privileges that came with my position as a high-ranking government official. In stark contrast, when my colleague Federico "Freddie" E. Puno and I went on a business trip to Japan in 1989 to negotiate with potential partners for Republic Glass, we were riding the bus! Since Freddie worked with me at PNOC and the National Power Corporation, he must have been struck by the drastic change in circumstance. When we arrived at Tokyo Imperial Hotel, he said, "Do you remember when we used to come here for official visits? Our hotel rooms looked like a funeral parlor because of all the floral arrangements."

But I was not the type who would wallow in self-pity. I had more pride than that. I kept to myself most of the time—playing golf, reviving my interest in playing cello, reading extensively, traveling with my wife and my two sons, and resuming my role in Republic Glass, a company where I was a majority owner. After the Sandiganbayan acquitted me in several criminal cases and subsequently lifted my travel restrictions in 1989, I made it a point to stay for three months in San Francisco every year. It was as though life was beginning anew for me, with old passions rekindled and new ones created. Yes, there were difficult moments, and there were times when I missed the old days in the Cabinet and the ministry, but there was enough for me to do to keep myself more than occupied. Principal among these was my return to private business.

Since I had been away from Republic Glass management from 1973 to 1986, I devoted myself to redirecting and reorganizing the company. I had professional managers to take care of the company in my absence, but unfortunately, none of them was dynamic enough to lead the company amidst new challenges and shifting business environments. They were also conservative in the sense that they were hesitant to confront the issues squarely, which hampered their decision making and inevitably affected the company's performance. When I returned to Republic Glass in 1987, I prioritized the modernization of the glass-manufacturing process, which had become so antiquated that we were producing glass of lesser quality. However, we could not afford the new technology available then, which cost about US$ 100 million, so we had to look for a partner. Fortunately, I knew the chairman of Asahi Glass of Japan over the years. I had been in communication with him, and as early as 1986 I had broached the possibility of having a joint venture between Republic Glass and Asahi

Glass. In 1988, he gave me a call, saying that they were ready for a joint venture. We sold 49 percent of the company to Asahi Glass. Subsequently, we jointly arranged a financing scheme of US$ 50 million from the Japan Export-Import Bank and Php 500 million from a consortium of local banks, to meet the financial requirements of building a new and modern glass factory. Ironically, that joint venture became the first major foreign investment under Mrs. Aquino's presidency.

Then I got involved in Air Liquide, a French company that is the largest supplier of industrial gas in the world, and which has a presence in every major area that needed a gas infrastructure. Air Liquide happened to supply Republic Glass with nitrogen and hydrogen, and even built the gas plant alongside the glass factory. In the late Eighties, company officials approached me for a possible joint venture that would service the emerging electronics industry, which was heavily dependent on gas for microprocessing. Republic Glass Holdings Corporation (RGHC), my holding company, took a 40 percent stake in the new company, called Air Liquide (Philippines). However, because of the constant demand for huge capital outlays, RGHC eventually sold out in 2000, although I am still part of the company's board of directors.

As for my personal life, it is fortunate that I had the foresight to return to cello playing in 1983 after twenty-seven years of not being able to play a single note. Thankfully, my good friends Oscar Yatco and Nena Villanueva came to my rescue. They showed inexhaustible patience by playing simple trio pieces just to get me going. When the pieces we were reading had consecutive sixteenth and thirty-second notes, I would rise to my feet and give up, but Oscar would "solfeggio" that portion for me very slowly, and I would be too embarrassed not to follow. But I managed to really brush up on my cello while in San Francisco, where there were many amateur cellists like me. I joined several nonprofessional instrumentalists to perform chamber music. I used to play with one who was a retired Justice of the Supreme Court of California; he played the viola, and we used to get together once or twice a week.

When I returned to Manila, I played chamber music with some members of the Manila Symphony Orchestra. They would come to my house and we would play for two or three hours. Later, I would go to St. Scholastica's College and play with some of the music students. I think what endeared me to these musicians was that I was one of the few who still played chamber music; not only that, I also have the most extensive library of chamber music that some schools borrow.

My return to cello playing also gave me and my friends Billy Manalo and Nena Villanueva reason to resume our trio, which we used to have in our younger days. For about three years, Billy and Nena came to my house once a week so we could play together. When Aling Buenaventura, the widow of National Artist Col. Antonino Buenaventura, heard about our weekly sessions, she joined us and we formed a regular quartet. I also performed with Oscar and Nena in several public concerts, especially for the annual Las Piñas Bamboo Organ Festival, where we played for a number of years.

To complement my musical interests, whenever I was in San Francisco I spent time around Silicon Valley visiting garage-type sound laboratories and shops specializing in audiovisual equipment. This encouraged me to build my own home-theater system in Manila, which I found to be relatively culturally barren since we could not afford the expense of bringing in major world-class performers.

Traveling was another regular feature of my life as a private citizen. Every year, besides staying in San Francisco for three months, I made it a point to spend at least one month in Europe with my wife, children and grandchildren, including at least two weeks in Scotland to play golf with my two sons. We did the latter for eight consecutive years (1992-2000), to the point that I received an invitation to join the Royal Dornoch Golf Club, an exclusive links course founded two hundred years ago in Dornoch, Scotland.

Over the years, I made new friends and acquaintances, some of whom were among my critics when I was still in government. I am also proud to say that all of my former PNOC colleagues continue to keep in touch, and whenever we have the chance to recall the past, all of us speak of PNOC with a sense of pride.

This book is my way of sharing a part of that past, confident in the thought that, ultimately, history will be the best judge of the efforts that we all make. That I made.

I may sound combative and irreverent at times, but I am no iconoclast. I hope that the readers will view this work with empathy and understanding. After all, it took me all of twenty years to be able to gather the proper time and historical perspective to reply to accusations of impropriety in governance and conduct leveled at me personally and the PNOC family.

This is my story.

Geronimo Z. Velasco's Letter of Resignation

Republic of the Philippines
Ministry of Energy

Geronimo Z. Velasco
Minister

July 1, 1985

His Excellency
President Ferdinand E. Marcos
Malacañang, Manila

Dear Mr. President:

I am writing the President regarding a series of news items published by the *San Jose (California) Mercury News* and which was the basis of an extensive report in a major local paper. As the President is aware, the report claims to be some kind of an investigative report on alleged foreign properties of prominent citizens but in reality was nothing more than a rehash of old gossips designed for obvious political end.

While this experience may not be uncommon to those actively involved in politics, the President knows only too well that I am not a politician, have never been one and entertain no intentions of becoming one in the future. The President will therefore understand why I have on several occasions expressed both my and my family's very grave concern on the humiliation and indignity that we are constantly being subjected to in the political shooting gallery.

I have, Mr. President, spent the last few days in deep soul-searching and I am convinced that we have not seen the end of this kind of virulence. Considering the temper of the times, I am sure this type of politically inspired journalism will not stop. It is to assume that truth and fair play will be the least of these journalists' concern.

Consider Mr. President the recent *Mercury News* "exposure." It condemns by innuendo. The readers of this one-sided type of reporting have no way of knowing that before I joined government, part of my compensation since 1965 to 1974, was a dollar component that was duly reflected in my yearly Income Tax returns. They also have not been told that I received normal employee stock options from the Castle and Cooke group of companies in view of my stewardship of their Philippine affiliates.

At the same time, the readers may never even know that I have never taken advantage of my government positions to obtain loans or guarantees from any government financial institution. People are also never told that when the President drafted me into the government services, I actually gave up a very lucrative position in a major multinational. I even recall, Mr. President, that you yourself mentioned my reputation as the highest-paid Filipino executive then. As is expected, people end up believing what they wish to believe and conveniently ignore the facts that contradict the conclusions they have already made.

All these Mr. President have been hard on me and my family. But more than just the personal dimension, I also feel that my usefulness to your government and to our people has been adversely eroded, the truth and a clear conscience notwithstanding. I have, Mr. President, always considered my government service the high point of my professional career. It has always been a distinct honor and privilege to be a part of your administration at such a critical time in our nation's history. Inasmuch as a professional manager's integrity is his most important management asset, the issues affecting my own integrity, no matter how unfair and unfounded, have raised questions as to my future ability to be effective.

In this regard, the President may wish to take note of my following concerns:

1. The Philippine National Oil Company enviably, since its inception, is the only government corporation that has enjoyed the highest financial credibility with international financial institutions. It has been able to borrow without government guarantees as well as deal with international crude suppliers, government or otherwise, on a footing of mutual respect.

   With these insinuations affecting my integrity (transcending the domestic press since the reports are also carried by the wire services), I feel that my continuance in office at PNOC may affect the very respect and esteem that these international institutions have for PNOC. I feel that I must now dissociate myself from PNOC in order to preserve PNOC's credibility because its continued success is so critical to the attainment of the objectives of government and for the good of our country and people.

2. I have endeavored to professionalize PNOC, the largest corporation in the Philippines, by imposing private-sector norms of discipline and performance on its officers and employees. Any deviation from this strict corporate code of behavior is immediately met with appropriate sanctions. In view of the cloud that now hovers over my person, I may no longer be able to continue to impose the discipline and office decorum on the staff, which will consequently affect PNOC'S future efficiency.

3. One of the most crucial issues presently engaging the public's attention, Mr. President, is the nuclear plant. I have been engaged in extensive and numerous speaking engagements to defend and justify the position of government on this issue. It is, however, clear that any defense of the nuclear program must be blessed with unimpeachable credibility, a quality which in the eyes of the public, I now feel I may no longer possess.

4. The President is well aware that it is the Board of Energy that regulates petroleum prices. Yet in the perception of the public, it is inescapable that they should associate me with these regulatory matters. These reports will now also affect the public's appreciation of the reasonableness of the prices of these consumer-sensitive products.

In view of all these, Mr. President, I wish to reiterate my long-standing request to retire from the public service. I realize only too well that I have promised the President that I will stay on until 1987. But I know that the President will also sympathize with the need to maintain the effectiveness of the institutions entrusted to my care as well as my need to vindicate my name and that of my family. As a professional manager, I am ill-equipped to fight a politician's battle. I find the respectability associated with public office a serious constraint that renders me incapable of taking issue with the fault finders of our times.

By requesting to be allowed to retire from public service, the President may accuse me of running away from a fight. Let me assure the President that as soon as my request to revert to private life is granted, I shall feel freer to mount a campaign to clear my name. Besides, Mr. President, it seems that in the context of today's political milieu, a professional manager and entrepreneur who is a non-politician, may no longer be as effective in government as in the past.

I have always tried to serve the President and the Republic to the best of my abilities. Even sectors that are normally critical of government projects have had to admit that the programs so graciously entrusted to me by the President have been among the best managed and in the case of corporations, most efficient and profitable. The mandate for greater energy self-reliance has likewise been achieved, with the 50% goal for domestic energy sourcing by the end of this year clearly attainable. The mission the President has entrusted to me has been largely accomplished, and the basis for future projects has been set.

As I reiterate my request to be allowed to return to private life, allow me to state that I will always be indebted to the President for the many opportunities and honors the President have bestowed upon my humble person. Let me assure the President of my loyalty and appreciation for the President's esteem and confidence.

Respectfully yours,

GERONIMO Z. VELASCO

CHAPTER ONE
CONFRONTING the OIL CRISIS

It was a bright Tuesday afternoon on December 18, 1973. Clad in black riding pants and a comfortable riding shirt, I was at the Republic Glass factory in Pasig, having a good time with Lightning, my favorite horse. Then, from the car's two-way radio came a message from my office: I had a call from Malacañang—President Marcos wanted to see me immediately. I looked for the nearest phone and called up the Palace. Upon confirming the message, I got into the car and ordered the driver to head for Malacañang. As soon as I arrived, security aides hurriedly escorted me to President Marcos's study room, where Cabinet members, ambassadors, some generals, and a few businessmen waited for him. Soon he entered the room and, upon seeing me, came over and said in quite a stern voice, "This time, Ronnie, I will not take 'No' for an answer." The next thing I knew, my good friend and Defense Secretary Juan Ponce Enrile took off his business coat for me to wear over my riding shirt and jodhpurs; aides instructed the television crew to take upper-body shots only, and then President Marcos swore me in as president of the Philippine National Oil Company.

President Ferdinand E. Marcos swearing in Mr. Geronimo Z. Velasco as president of the Philippine National Oil Company. Note the absence of Sec. Juan Ponce Enrile's coat, as it was loaned to Velasco for the occasion. (December 18, 1973)

GZV meets Crown Prince Fahad of Saudi Arabia (right), with Chief Royal Protocol Officer Dr. El Keireji. (June 1975)

Nineteen seventy-three was a crisis year not only for the Philippines, but for many countries in the world. It was the time of the first "oil shock," when oil prices soared, production supply decreased, and oil-importing countries faced the nightmare of gas rationing, rising oil import bills, and the slowing down of economic activities that depended heavily on oil and oil-based products.

The event that triggered the oil shock was the decision of the Organization of Petroleum Exporting Countries (OPEC), led by Saudi Arabia, to raise oil prices, cut back on production, and impose an embargo on all oil shipments to the United States (US) and other countries that supported Israel in the Arab-Israeli war, which had erupted on October 6 of the same year.[2]

Events unfolded rapidly in that fateful month. Immediately after Egypt and Syria launched a joint attack on Israel, the government of Iraq announced that it had taken over Mobil and Exxon's oil interests in the country.[3] By the third week of October, six Persian Gulf nations announced an increase of 21 percent in oil prices—from US$ 3.01 to US$ 3.65 per barrel—while the Arab nations decided to cut back on oil production by five percent every month "to bring pressure on Israel and the countries that supported her."[4] Libya delivered its own double whammy by increasing prices from US$ 4.604 to US$ 8.925 a barrel, and imposing a total embargo on oil shipments to the United States.[5]

If we look back, the oil shock of 1973-1974 should not have surprised world leaders and oil-importing countries at all. The power that comes with the possession of oil began to manifest as early as September 1960, with the formation of OPEC. Led by the top oil-producing countries—

---

2. "MAN OF THE YEAR – Faisal and Oil: Driving Toward a New World Order," *Time Magazine*, January 6, 1975 (Cover Story).
3. "Peace shattered and oily diplomacy," *New York Times*, October 8, 1973.
4. "Japan is stunned by Arab oil cuts," *New York Times*, October 19, 1973.
5. "The Arab oil weapon comes into play," *New York Times*, October 21, 1973.

Saudi Arabia, Iran, Iraq, Kuwait, and Venezuela—OPEC was subsequently joined by Abu Dhabi, Qatar, Libya, Nigeria, Algeria, and Indonesia. Together, these countries made up more than 50 percent of world oil production and nearly 90 percent of oil exports. By creating an organization that would unilaterally determine the production and pricing of oil, the OPEC countries signaled to the rest of the world that oil would no longer be obtained as freely and as cheaply as the American and European oil companies used to do.

Nonetheless, the intimate connection between oil and politics became obvious only in the latter half of the Sixties, in relation to several events. The Arab-Israeli war of 1967 forced the closure of the Suez Canal, which strained supplies from the Persian Gulf and sent tanker rates soaring. A civil war in Nigeria cut off a major oil source. In 1969, the overthrow of Libya's King Idris by a young colonel named Muammar al Qaddafi resulted in an aggressive new government that successfully obtained higher payments from oil companies working in Libya's vast fields. OPEC, led by Saudi Arabia and Iran, followed Libya's example by agreeing to raise prices during the 1971 Teheran conference.[6] Throughout these events, OPEC members debated the appropriate oil policy that would best advance their respective interests. Some favored *participation*, in which Western oil companies would give up a share of their concessions to the host country while retaining operating control; others favored *nationalization*, which entailed the seizure of ownership or control by the national government.[7]

Taken together, these developments showed that the Persian Gulf nations were no longer the pliable states that the Western powers (especially the former colonial overlords in the Middle East) and oil companies had taken them to be. It is therefore not surprising that, on the eve of October 1973, some quarters had already been warning about the Arabs wielding the "oil weapon" to obtain economic and political concessions. But the "oil shock" nevertheless stunned the world not only because of the rapidity in which OPEC raised oil prices, but also because,

---

6. "Energy Crisis: Shortages Amid Plenty," *New York Times*, April 17, 1973.
7. "Oil: Participation vs. Nationalization," *New York Times*, October 22, 1972.

as many observers noted at the time, the US, the United Kingdom (UK) and other major industrial powers underestimated the organization's willingness to use the oil weapon as a bargaining chip. As the Shah of Iran proclaimed at the time: "The era of terrific progress and even more terrific income and wealth based on cheap oil is finished."[8]

The events in the Middle East severely affected us since our country's importation of crude oil depended completely on US and UK-owned oil companies. We can say that the local oil sector represented an accurate snapshot of the domination by American and other foreign companies of key economic sectors. In 1970, there were six major oil companies engaged in the marketing of oil products and the operation of four refineries: Shell, Caltex, Esso, Mobil, Filoil Refinery (a Gulf Oil affiliate), and Getty. Filoil Marketing Corporation, one of the local petroleum resellers, had a Filipino component but the US firm Gulf Oil effectively controlled the organization. Our heavy dependence on foreign oil companies was aggravated by the fact that the Philippine government, prior to 1973, had no agency for oil procurement, limited knowledge in negotiating oil agreements, and no domestically owned oil tankers that would ship crude from overseas.

The country was in a precarious situation. The year past, 1972, was an eventful one. There was Typhoon Yoling, which devastated Central Luzon and set back our economy. Apart from flooding thousands of hectares planted to rice, corn, sugar and vegetables, the typhoon caused heavy damage to infrastructure, telecommunications, and private property. Such natural disasters were worsened by man-made ones.

The decade of the Seventies began in political turmoil. Student demonstrations constantly rocked the capital, the rivalry between the president and some political interests became dangerously bitter, and unrest appeared to spread in Muslim Mindanao. President Marcos responded by declaring martial law on September 21, 1972. For someone who felt alienated by the sight of daily rallies and the general state of lawlessness, I personally welcomed the president's action. Although it

---

8. "MAN OF THE YEAR – Faisal and Oil: Driving toward a New World Order," *Time Magazine*, January 6, 1975.

ended almost thirty years of American-style democracy, I thought martial law put order where there was none.

On top of our political problems, the local oil industry joined the fray, so to speak, to add to our miseries. Sometime in 1973, Gulf Oil finally decided to pull out of the Philippines and sold its majority and controlling shares in Filoil to the National Investment and Development Company, the investment arm of the Philippine National Bank. Then Esso (Philippines) Inc. followed suit. Discouraged by the increasing regulation of the industry,[9] the parent company Esso decided to fold up its Philippine operations at the height of the oil crisis in December 1973, selling everything lock, stock, and barrel to the government.[10] Esso provided a transition crude supply contract for a limited time that would allow the government some breathing spell in the aftermath of their pullout. However, since both Esso and Filoil accounted for approximately 35 percent of our oil market, their departure was expected to create a major shortfall in oil supply once the transition crude supplies ran out. At a time when 97 percent of our energy needs came from imported oil (roughly 80 percent of which was provided by Saudi Arabia and Kuwait), the pullout of the two oil majors posed potentially huge problems that the country could ill-afford.

This is not to say that the government watched idly as these events unfolded. On the contrary, President Marcos had already made several moves in anticipation of a global energy crisis. On December 31, 1972, he issued Presidential Decree (PD) 87, otherwise known as the "Oil Exploration and Development Act of 1972," which mandated a more aggressive and active government participation in the actual search for indigenous oil resources, through the extension of incentives and liberal contract terms to financially qualified and technically able foreign and local oil exploration companies. To attract investors, PD 87 introduced

---

9. The Oil Industry Commission set the average pump price of fuel, for example. Oil companies also had to secure the Commission's approval before they could import petroleum products.

10. Getty would later sell its Philippine holdings to Pilipinas Shell in 1980, while Mobil would pull out in 1983, thereby leaving the Philippine market to the control of three refineries—Shell, Caltex and Petron/PNOC.

the *service contract system*, replacing the old but ineffective concession system.

The concession system for oil was similar to that for mining: Once a concession area had been identified for exploration, the private investor would submit to the government his estimates of the expected volume of oil that could be tapped and the costs to be incurred in exploration. After selling the oil in the international market, the investor would recover his costs and then take his share of the profit, with the remainder paid to the government. Now, the problem was that given the very complicated accounting system, oil concessionaires usually had a lot of leeway to pad costs, even to the point of claiming that they lost money, so that the government seldom received anything from whatever oil was sold by these concessionaires.

The service contract system, on the other hand, was based on the concept of production sharing, which upheld the sovereignty of the producer-country over its own resources and guaranteed a share of the product itself without the royalties and complicated accounting of the old concession system. At the same time, fair returns awaited firms (as service contractors) that assumed oil exploration risks and provided the necessary technology and capital. For exploration ventures that the Philippine government financed, a stipulated service fee was paid to the service contractor for having furnished services and technology. On the other hand, if the service contractor financed the exploration besides providing services and technology, the company's operating expenses would be reimbursed besides being paid a service fee.[11] Caltex, through its international shareholders Chevron and Texaco, signed the first service contract with the government in 1973 to explore oil in Philippine waters.

Another major action by the President was the issuance of PD 334 ("Charter of the Philippine National Oil Company") on November 9, 1973. To my mind, it was this law that really set the Philippines on the road to addressing the oil crisis systematically, holistically, and with a long-term

---

11. Presidential Decree 87, "The Oil Exploration and Development Act of 1972," December 31, 1972. See in particular Sec. 6 and Sec. 7 of the said decree.

perspective. I am emphasizing this because, from my experience, I can attest that President Marcos really had a strategic vision insofar as energy development was concerned. Instead of tackling issues like oil price hikes on a piece-meal basis, which seemed to be the tendency of succeeding political leaders, President Marcos sought to address these matters in the context of long-term energy goals.

To the cynical or the unbelieving, PD 334's Declaration of Policy—that the state's pursuit of industrial and overall economic development was premised on the "effective and efficient utilization of energy sources"—is no different from the usual "motherhood statements" one finds in the basic document of any organization. But President Marcos went beyond platitudes. He recognized that the government needed to build its organizational and technical capability to match his vision of energy self-sufficiency. He also realized the necessity of minimizing our country's dependence on the foreign oil companies, if we were to reduce our vulnerability to the vagaries of the world oil market. He created the Philippine National Oil Company (PNOC) as the organizational vehicle that would ensure the adequate supply of oil and oil products, as well as power and energy, to all users.[12]

But PNOC remained nonoperational for a month, since it was not until December 12, 1973 (barely a week before my appointment), that its Board of Directors was constituted. President Marcos appointed Vicente T. Paterno, secretary for industries, as chairman and acting president, with the following as directors: Juan Ponce Enrile (defense secretary), Alejandro Melchor (executive secretary), Cesar Virata (finance secretary), Estelito Mendoza (solicitor general), Panfilo O. Domingo (president, Philippine National Bank), and Leonides Virata (chairman, Development Bank of the Philippines).[13]

As an immediate response to the uncertainty in supply caused by OPEC's cutbacks in production as well as the departure of Gulf Oil and

---

12. Presidential Decree 334, "Charter of the Philippine National Oil Company," November 9, 1973.
13. "President names PNOC Board," *Philippines Daily Express*, December 12, 1973.

Esso from the Philippines, the government imposed gas rationing. The remaining oil majors—Shell, Caltex, Getty, and Mobil—could not fill the vacuum hitherto filled by Gulf Oil and Esso, since their mother companies refused to sell additional crude oil for the Philippine market. Such uncertainty compelled President Marcos to issue General Order 41 on December 10, 1973, directing the PNOC to assume supervision over the sale and distribution of all available stocks of crude oil and oil products, whether imported or produced by the local oil refineries.[14] General Order 41 technically put the oil industry under the government's control and provided the latter with the legal basis for requiring oil companies to submit inventory reports regularly. These reports became the government's basis for determining whether it should impose rationing.

Simultaneous with its intervention in the domestic oil market, the government also prepared to take over Esso's local subsidiary. Considering that it had already controlled Filoil through the Philippine National Bank, the government's purchase of Esso (Philippines) enhanced its presence in the industry.

This was the situation when I entered government.

Prior to that fateful day in December 1973, I already had twenty-two years of experience in the private sector. I had a brief stint in the marketing group of an oil company, Standard Vacuum (Stanvac), in 1951-53.[15] I was not happy with the job and so, in 1953, I moved on to Borromeo and Lim Management Consultants, organized by my friends Victor Lim and Federico Borromeo. It was in this firm that I had my first encounter with a real engineering job, when I was assigned to establish a porcelain enamel plant. Following the plant's completion in 1958, I decided that I needed a new challenge.

One day, while reading a newspaper's business section, an item caught my eye: Harry Stonehill was putting up a glass plant. Knowing that a glass plant needed a furnace, I thought maybe I could build it for them. The news item also said that Stonehill had just organized Republic Glass and

---

14. "Government in total control of oil," *Bulletin Today*, December 11, 1973.
15. Stanvac later became Standard Oil, better known as Esso.

applied for incorporation with the Securities and Exchange Commission, and that among the incorporators was Leonides "Leo" Virata. Since I knew Leo, I called him up and offered my services. He agreed to arrange an appointment with Stonehill, who interviewed me a week later and hired me right there and then. The following Saturday afternoon, he brought me to the "glass plant"; it turned out to be an open field in Pasig used as duck farm for producing *balut*.[16] This was where I would *build* the plant from scratch, in 1958.

It took two years to build the glass plant, the management of which Stonehill entrusted to me completely. There were not many Filipino managers at the time and most of Stonehill's operations were run by foreigners, but probably he got the right vibrations for me. He involved me in several other projects such as the Atlas Cement (forerunner of one of the cement companies that grew in the 1970s), the design and construction of the Manila Memorial Park, a reclamation project, and his cigarette company.

I enjoyed my work at Republic Glass, fulfilled in the thought that I was applying my knowledge as a mechanical engineer and testing the limits of my capacity. Then came a turning point in my career. I was already general manager when, one day in March 1962, Harry Stonehill was arrested and made to undergo deportation proceedings.[17] A week later, I woke up one morning and was told, "You are now president of Republic Glass."

How did it happen? The board members of Republic Glass were Jose "Jobo" Fernandez, Carlos Palanca, Leonides Virata, Romeo Villonco, and Esperanza Zamora—fairly well-known people in that business era. Not one of them wanted to succeed Stonehill and take over the company. Stonehill himself provided the solution to the impasse. He told them, "Okay, if you leave it to me, I'll take Velasco. Let's put up Velasco, allow

---

16. A Filipino delicacy of duck embryo.
17. The so-called "Stonehill scandal" dominated the headlines of major Philippine newspapers in 1962, when then-Justice Secretary Jose W. Diokno investigated Harry Stonehill for alleged bribery of high government officials and politicians. President Diosdado Macapagal subsequently ordered Stonehill's arrest in March 1962, and his immediate deportation in August of the same year.

him to take my seat in the board, and you can elect him president." The next day, as scripted, I was elected president of Republic Glass. That was in March 1962, and I was thirty-four years old. I am proud to say that, in the year I became president of Republic Glass, the company started to make money for the first time since it began operations. From 1962 to 1972, we enjoyed annual rates of return that averaged 20 percent, making the company very profitable.

Because of his troubles in the Philippines, Stonehill sold his shares in Republic Glass to Castle and Cooke Inc. of Honolulu, Hawaii, in August 1963. I had not even heard of Castle and Cooke, and yet I was being told that this company would become my new employer. It turned out that Castle and Cooke owned Dole (Philippines), which was the single largest foreign investment in the country at the time, worth US$ 45 million. In 1965, Castle and Cooke appointed me as vice president and, two years later, president of Dole (Philippines). This was how I earned the reputation of being the country's highest-paid executive, whether Filipino or foreign, at age thirty-nine.

Such was my personal circumstance in December 1973, when, without warning, President Marcos appointed me as PNOC's chief executive officer (I found out later that Juan Ponce Enrile mentioned my name to the president). On my way to Malacañang that afternoon, I already had a nagging feeling that the president would offer me a position, and I was uncertain whether this time I could afford to decline. Twice he had offered me a post in his administration: in 1966, he asked me to head the Bureau of Customs and, in 1970, he offered the job of commissioner of the Bureau of Internal Revenue. Both offers I had managed to decline gracefully, citing professional commitments as well as lack of expertise (especially in matters of taxation). Thus, when he said on that fateful day, "This time, Ronnie, I will not take 'No' for an answer. You're going to head a national oil company," I knew he meant it. I did not even get the chance to call up my wife or my employers in Hawaii to tell them that I was getting appointed to a government job. After I was sworn in, I asked, "Mr. President, how long will I stay in this post? What should I tell my wife and my employers?" He answered, "Oh, tell your wife, four years." That was all he said. Things happened so fast I did not have time to think about the drastic fall in income that would come with the PNOC post. As the late Teodoro "Ka Doroy" Valencia wrote the next day in his newspaper column, I was "giving up gross earnings of at least Php 1 million for the Php 36,000-a-year position."[18]

Why did the president choose me? Certainly, my past exposure to him was a factor. When I was president of Dole, and whenever he had the opportunity to go to Mindanao, President Marcos stayed in the company headquarters in South Cotabato. In fact, he visited us several times. After going to various places, he wound up sleeping at Dole, often for several nights. He would live in my house, which I would vacate and turn over for his use. He liked the weather in South Cotabato, not to mention that my house was built practically in the middle of a golf course, so he spent a lot of time at Dole and that was how I got to know him. Later, we would play golf at Wack-Wack and on the Palace greens. Yet, my relationship with him remained at the personal level, since I never really had significant transactions with the government.

Outside of personal acquaintance, I would like to think that the president chose me because of my exposure and experience in corporate management. Besides heading Castle and Cooke's Philippine operations as president of Dole (pineapple), Standard Fruits (banana), and Republic Glass (glass), Castle and Cooke made me oversee their operations in Thailand (water pipe company) and Singapore (ready-mix concrete aggregate company) as chairman of both companies. Maybe, since President Marcos intended PNOC to operate as a corporation, he thought that I could steer the state-owned company in the same manner that I managed these private corporations.

I can only speculate as to the President's reasons for appointing me. All I know is that, then-Solicitor General Estelito Mendoza told me, "This is the first time I've seen the President act without hesitation on an appointment—'*Ipatawag mo si Ronnie*' (Call in Ronnie)."

---

18. Teodoro Valencia, "Over A Cup of Coffee: The Right Man," *Bulletin Today*, December 19, 1973.

I myself had no problem working for President Marcos. Notwithstanding my past refusal to take up his offers of position, I believed strongly in his capacity for governance. I admired his intelligence, especially his grasp of the issues that confronted the country and his presidency, and I believed that he did the right thing in declaring martial law. At the same time, although I was not as close to him as others had been, I had developed sufficient rapport with the president in the course of hosting him at Dole-South Cotabato and playing golf together.

If I had any personal concern about taking on the PNOC job, it had to do with the prospect of encountering all types of hangers-on and favor-seekers whose only claim to legitimacy was their closeness to the First Couple. Modesty aside, I was the wealthiest political appointee at the time—I was even ranked ahead of President Marcos in the country's list of Top 100 taxpayers in 1974—and I did not want anyone taking advantage of my position.

So, at the onset of my PNOC stint, I requested the president to sign a letter which I prepared and which stated that my authority to make donations on behalf of PNOC was limited to Php 5,000 per donation, and that any amount in excess necessitated prior approval by the president of the Philippines. I made that move to protect myself professionally and personally. The president understood and readily signed the letter. Over the years, and as I had expected, this letter became handy on occasions when friends and associates of the First Couple approached me for contributions. They backed off the minute I waved the letter containing the president's signature. If anyone claimed to have President Marcos's blessing, I responded that I still needed to secure the president's written instructions; the request would usually be withdrawn.

I did have some reservations about heading a national oil company. Frankly, I knew nothing about oil and energy. But then again, neither did I have any prior expertise in glass manufacturing, pineapple, and banana—fields that I entered in the course of my professional life. What carried me through, I think, was the fact that I was always prepared to learn, and I never lost the interest to learn more, notwithstanding my position at the top of the company. This attitude helped me a lot in adjusting to my new role at PNOC.

Just the same, my appointment as PNOC president brought on a different kind of pressure since my new job directly impinged on the national interest, as the president himself emphasized. As mandated by PD 334, the PNOC's main objectives were:[19]

- To provide and maintain an adequate and stable supply of oil and petroleum products for the domestic requirements;
- To promote the exploration, exploitation and development of local oil and petroleum sources; and
- To foster oil or petroleum operation conditions conducive to a balanced and sustainable growth of the economy.

Given the prevailing conditions, however, PNOC's immediate task was to secure a stable and affordable supply of oil that would cover Esso (Philippines) and Filoil's share of roughly 35 percent of the local market. In fact, the president's first instruction was for me to go to the Middle East to forge government-to-government relations in the procurement of oil supply. I remember, after my swearing-in, the president kept saying "Riyadh—go to Riyadh," and I did not know who or what on earth Riyadh was! At some point I had to interrupt him to ask what he meant by "Riyadh." You can imagine my embarrassment when he said that Riyadh was the capital of Saudi Arabia.

Now, you may ask, why did President Marcos insist that we buy oil from governments? Well, he believed that obtaining oil on a strictly commercial basis would render the country vulnerable to the uncertainties in the world oil market. By contrast, government-to-government relationships appeared to be more reliable and easier to manage in the sense that there was more transparency in the negotiation. There seems to be an unwritten rule or practice that government-to-government contracts are much more solid or well-grounded since it is the government

---

19. Presidential Decree 334 (Creating the Philippine National Oil Company).

you are talking to. Of course, the oil companies that remained in the Philippines—Shell, Caltex, Mobil, Getty—had their own suppliers and purchased crude oil on a commercial basis. So my job was initially limited to covering the supply of Esso (Philippines) and Filoil.

I had barely warmed my seat as PNOC president when the government formalized its acquisition of Esso (Philippines) on December 21, 1973. I signed the contract on behalf of the government, with T.E. Wallace of Esso Eastern as my counterpart. For a cash payment of US$ 19.5 million, the government obtained twenty-eight bulk plants, more than 800 stations that serviced Luzon and Southern Philippines, and a 57 percent share of the Bataan Refining Corporation (Mobil Oil owned the remaining 43 percent). The deal also provided for an Esso affiliate to provide PNOC with transition crude supply and for the service stations to continue using Esso's name for sixty days. On the same occasion, the company was renamed Petrophil Inc. The purchase was, as the newspapers headlined, "a milestone" for the government. Nonetheless, President Marcos made the effort to clarify that the government would sell its holdings in Esso (Philippines) "as soon as private Filipino enterprises have the funds and capability to take over."[20] It was as if the president wanted to allay the fears of foreign companies about the possible nationalization of the oil industry.

Obviously, things were happening fast and I felt pressured to keep in step. Although I understood the urgency of ensuring continuity of oil supply, I really had no idea about the scope and depth of the problem. The first thing I did was to learn what the oil business was all about. I needed an education because, on top of the things mandated by PD 334, I was about to take responsibility for a petroleum marketing organization (Petrophil) and two oil refineries (the Bataan Refining Corporation and Gulf Oil's Filoil Refinery) that together accounted for more than a third of the country's total capacity.

I spent the remaining days of December 1973 trying to learn as much as I could about my new job. One of the first persons I called for to give

---

20. "Esso deal: Milestone for RP," *Philippine Evening Express*, December 22, 1973. See also "Govt acquires Esso (Phil.), Inc." by the *Times Journal*, published on the same date.

me a situationer was Antonio "Tony" V. del Rosario, who was vice president for corporate planning and finance at Filoil. He had done a lot of traveling to the Middle East on behalf of Filoil (which by then was already government-owned and -controlled), trying to secure supply contracts since the transition supply contract from Gulf Oil was expiring.

I met Tony a few days before Christmas. He acquainted me with the operations of Filoil and the efforts they made in augmenting the company's crude supply. After my briefing session with Tony, I talked to other top executives who handled different aspects of Esso's and Filoil's operations. In relation to crude supply, one of the first things I learned was that I could not just buy any oil—that the Bataan refinery could only process a specific kind of oil for which the refinery was designed. That was the first time I heard that there were different types of crude oil, and that refineries didn't just process any oil.

Then I had to learn about the petroleum marketing system. Looking back, there is nothing very intricate about the marketing needs for petroleum. At the time, though, I needed briefing because if you were in the business of marketing petroleum products, you had to have products all over the country. To be able to do this, you had to have bulk plants where you can store your product and where the outlying areas can draw their fuel. Now, how do you supply these various bulk plants? There has to be a logistical support system in terms of barges, large tank trucks, and other vessels to move the product. Those were all learning experiences for me.

Now, I was fortunate that I did not have to organize those details because they were already in place. All I did was to familiarize myself— "*O, ganun pala ginagawa* (So that's how things are done)." It was for me to discover, to be told that, "This is the way we do things." I was not about to change anything because this system was designed by Esso based on US standards, and these guys who briefed me were experienced personnel in the field. Many of them were US-educated, some had PhDs from US universities, and all had years of experience as well as additional training in the course of working for the oil companies. I was confident that these people, whom I simply inherited, were competent for the job.

But I am not saying that my entry into PNOC was completely without any hitch. In any organization, the arrival of a new boss is often met with anxiety as well as skepticism over the new chief's capabilities. Being a "new kid on the block," so to speak, I expected that there were those among the "big boys" of Esso (Philippines) and Filoil who doubted my capabilities. I also did not discount the fact that, compared to them, I only had a bachelor's degree in mechanical engineering from the "other MIT."

For my part, I had made it a matter of personal policy not to bring my own retinue of assistants into a new company. In my years as chief executive officer, I always tried to work with the existing personnel. That was why the only person I brought along to PNOC was my driver. But that did not mean that I was about to play the role of a new guy who would try to get into the good graces of his employees. On the contrary, I made a conscious effort to show them that, even if my knowledge of the oil business was limited, I knew how to manage a corporation. Since PNOC needed to complete the process of taking over Esso/Petrophil's and Filoil's refining and marketing operations, major organizational processes had to be undertaken. I started moving people around, making them take on new assignments and forcing those who could not work together to join the same team. As Tony del Rosario later recalled to a former colleague, "I guess you can describe that initial process as one of establishing a chain of command, and the first major impression I got of Ronnie was that this guy certainly knew how to establish a chain of command. He left no doubt in our minds as to who was the boss."

I also anticipated that, with the eventual merger of Petrophil and Filoil, PNOC would become an unnecessarily large company. Thus, we prepared a redundancy program that offered each employee the equivalent of two months' salary for every year of service. At the same time, I made sure that the key management executives would stay with me. I talked to each of them and asked if they could remain with PNOC and try out my leadership at least for six months; in return, I promised to keep their salaries and benefits intact. You see, most of these executives from Esso Philippines enjoyed much higher salaries compared to their Filoil counterparts, so I had to provide substantial incentives to make them stay in a government-owned corporation. That was the first thing I cleared with the PNOC board: "We can only keep these people if we keep everything intact."

A major challenge in merging Filoil and Petrophil was the integration of their refineries. Besides being located at a considerable distance from each other (Filoil Refinery was in Rosario, Cavite, while Bataan Refinery was in Limay, Bataan), the two refineries had a substantial difference in capacity: Bataan refinery had a rated capacity of 130,000 barrels per stream day, Filoil had only 25,000 barrels per stream day. There was no question that Filoil had to be sacrificed so that we could come up with a more efficient system. I assigned the respective vice presidents of the two refineries, Jose "Jovy" U. Jovellanos of Filoil and Jesus "Jess" Dulce of Bataan Refining, to conduct a study on how to integrate the two refineries, a task which they completed swiftly as both were very able managers. Integrating the Petrophil and Filoil refineries was more difficult than merging the two companies' marketing groups, and yet the process went smoothly under the efficient and able management of Jess Dulce[21] such that Bataan Refinery did not lose a single day of production. The integration of the two refineries eventually resulted in net savings on the cost of refining Petrophil's oil products. I remember our partners at Bataan Refining Corporation then, Mobil Oil, praising me for having made such a decision.

From the beginning of my term at PNOC, I was intent on reversing the traditional attitude toward organizational structures in government. I knew that the bureaucracy had a natural inclination to grow by itself and thus burden performance over time. So I made sure that we were going to be a flexible and effective vehicle for executing corporate strategies and achieving stated performance targets. Besides the usual corporate systems, financial resources and authority, what I needed most was an adequate human resource base. Given my experience in the private sector, I negotiated with President Marcos that PNOC be exempted from certain government regulations: No cut in salaries or benefits, no civil service

---

21. Jess Dulce had been trained extensively by Esso, having assigned him to refineries in Australia and Europe, as well as to the Esso Eastern headquarters in New York City.

rules to constrain our employment system, the installation of a performance-linked evaluation system for employee appraisal and advancement, and no government audit to slow us down. I explained to the president that, among others, clearing with the Commission on Audit (COA) the price of crude that I could obtain from the various supply countries was highly impractical. Besides, COA did not have the ability to decipher what the proper value of crude was, especially in a market that was ever so fluid. The president saw the validity of my proposal. Finally, I promised the senior executives that I would give them authority just as I would provide them with responsibilities, and that around them we would build the organizational structures that would facilitate our capability upgrading.

Personally, I view PNOC's exemptions from some government rules as a clear signal from President Marcos that he trusted me fully. Even if my position at PNOC was a non-Cabinet rank, the president gave me wide latitude in running the organization, vested so much authority in me, and allowed me unhampered access to him. I was one of the few who could call him up or see him without prior appointment, or even brief him while he took lunch, which he normally did not allow. Even my relationship with the PNOC board reflected my unusual position. Considering that the board had as members some of the country's most powerful men, such as Defense Secretary Juan Ponce Enrile and Executive Secretary Alejandro Melchor, I never felt that they doubted my capacity to run the company.

Important as organizational matters were, addressing them was just one of the things I had to do in my early days at PNOC. I still had to meet the president's demand that I obtain crude supply to cover the 35 percent share of Esso (Philippines) and Filoil. An unpleasant New Year's Day gift reminded me of the urgency of the task—crude oil prices jumped from US$ 3 to US$ 11 per barrel on January 1, 1974. I had to continue with the gas rationing that the government began earlier.

In those days, we required each of the oil companies to have a minimum inventory of sixty days' supply of crude and oil products—gasoline, diesel, kerosene, etc. Our rule of thumb was, if the total country inventory is less than sixty days but not lower than forty-five days, we ration. If stocks go below forty-five days but higher than thirty, we cut the ration in half. If the inventory comes down to thirty days or lower, we declare an emergency; only hospitals, power-generation systems, and other critical areas, would be allowed to operate.

When the oil price soared to US$ 11 per barrel, we had another shortage, so we had to ration for about two months. Thankfully, it was an orderly process—maybe because we commissioned a military general to administer the system. We handed out coupons indicating specific amounts of gasoline to car owners and other transport operators. Private car owners could only avail themselves of 200 liters a month but we gave more to the jeepney drivers, who needed thirty liters a day. The problem, as we found out later, was that the jeepney drivers sold their extra coupons! Since the coupons were color-coded to distinguish private cars from public utility vehicles, what these clever drivers did was to siphon off the fuel from their gas tanks, sell it in tin cans or plastic containers, and then line up again at the gasoline station to exchange the next coupon. No, I did not bother to send anyone to jail for economic sabotage. But that was when I realized that rationing, even under the strictest conditions of martial law, could not be enforced without other people taking advantage of loopholes in the system. Nonetheless, we continued to ration for about two months and lifted it after we secured our first crude supply contract.

By the first week of February 1974, I began traveling all over the Middle East, asking for supply of oil on a government-to-government procurement basis, with Tony del Rosario providing valuable assistance in acquainting me with the processes and issues related to these negotiations. All this time President Marcos, who had made me ambassador-at-large to facilitate my introduction to oil ministers and heads of state-owned oil companies, phoned me every so often. He kept asking, "What's happening? When are we getting a contract?" Finally, in late February 1974, we successfully negotiated our first government-to-government crude supply contract with Kuwait, which agreed to sell to Filoil 30,000 barrels of crude daily for six months, thereby helping to cover Filoil's deficit.[22] That contract gave me

---

22. "Kuwait sells crude to Filoil; deficit covered," *Philippines Daily Express*, March 1, 1974.

the confidence to do my job, and the experience served as my guide for subsequent negotiations. As we became exposed to the vagaries of the oil trade, we obtained similar contracts from Saudi Arabia, Indonesia and the United Arab Emirates.

On March 4, 1974, President Marcos appointed me as chairman of PNOC, besides being its president, in order to consolidate operations and to define my role better. Equally important, the Kuwait deal eased whatever doubts the Petrophil executives had about me; they saw that I could deliver on expectations, even if I knew nothing about oil. That certainly made it easier for me to exercise leadership once the merged Petrophil-Filoil came under PNOC management in April and I had to institute a top-level reorganization.

The formal merger of Petrophil and Filoil took effect on April 1, 1974, with Petrophil, being the dominant company, absorbing Filoil's operations. Another result of the merger was that the two companies' service stations changed their names from Esso and Filoil, respectively, to Petron.[23] The name "Petron" was actually the winning entry in a contest I held among the PNOC staff. Later, when I informed Malacañang that Petron would be the brand name for Petrophil products, First Lady Imelda Marcos teased me, "*Salbahe ka talaga! 'Petrolyo ni Ronnie' ang ibig sabihin niyan eh!* (You're such a rascal! That name stands for 'Petroleum of Ronnie'!)"

I continued traveling to the Gulf to secure more government-to-government contracts. I learned much in those trips to the Middle East. In the beginning, I had somewhat naively thought that it would not be so difficult to convince these oil-producing countries to sell affordable crude to us. After all, we were a poor Third World country and we needed help in getting our process of economic development started. Little did I know that other countries, rich and poor alike, were doing the same thing, wooing the OPEC countries with their own reasons for obtaining crude oil.

One encounter I cannot forget was my meeting with the head of the Iranian National Oil Company. His office was full of foreigners when we

---

23. "Petrophil, Filoil merge," *Bulletin Today,* March 28, 1974.

arrived. As soon as we sat down to talk, he said, pointing to the French delegation, "You see all those people? Well, they just offered to finance some of our projects. What have you got to offer? If you have nothing to offer, you're going to waste my time." So I went into a song-and-dance about the Philippines being a poor country and how we were just trying to establish ourselves, that we needed some oil to jumpstart economic activities, etc. The Iranian official cut me off—"You're just wasting my time"—and ended the meeting right there and then. That was a major lesson in oil-power politics.

Another revelation was the extent to which other countries came up with all kinds of devices to obtain access to powerful oil ministers. I cannot forget how the Brazilian delegation brought along the world-famous football player Pelé in their visits to the Middle East and made him play football with the children of the elite. I think my jaw dropped when I saw how Pelé worked his magic; he certainly aided Brazil in obtaining generous terms from the oil-producing countries.

As I became more familiar with the Middle East governments, I realized that we had to be prepared to offer something in return for whatever crude supply they could give us. The approach differed depending on the country. For example, with Saudi Arabia, we thought of offering our skilled Filipino laborers as contract workers. At the time, thanks to the huge inflow of petrodollars, the kingdom was beginning its construction binge for infrastructure and commercial facilities. But because Saudi Arabia lacked for labor, we thought that our Filipino workers could help fill the gap. The Saudi government welcomed the offer, and soon we were sending skilled workers over. In February 1974, when I made my first trip to Riyadh, there were only seventeen Filipinos in the kingdom, most of them being former hydraulic engineers from the Department of Agriculture. By 1978, the number of Filipinos in Saudi Arabia had ballooned to 300,000.

In relation, I recall that then-Labor Secretary Blas Ople and I had a lot of discussion about what kind of Filipino workers we should deploy for jobs overseas. At the time, in the mid-Seventies, the Department of Labor was already contemplating such a policy. I argued with Blas that we should only send our skilled workers, not domestic helpers. Why? Because I had a sense that sending domestic helpers—majority of whom were women—to foreign countries might just create more problems for us. I don't know, but somehow I felt that if many Filipinos could abuse their own domestic helpers, what more if foreigners—who are culturally different from us—employed them? President Marcos used to listen to us discussing the fine points of overseas employment. Ultimately, Blas prevailed. As he told me half-jokingly, "Look, Ronnie, you've had a comfortable life. You have no idea how desperate our jobless Filipinos are. If we don't provide them a safety valve like overseas employment, we might have a bigger problem in our hands!"

At this point, I also want to say something about how the secessionist problem in Mindanao affected our relations with Islamic oil-producing countries. Some books on the Moro National Liberation Front (MNLF) talk about how "oil politics" enabled key OPEC members to play an influential role in the resolution of the Mindanao conflict.[24] However, in my personal experience, the said conflict seldom figured in negotiations for government-to-government crude supply contracts. Of course, there were occasions when questions about the "Mindanao problem" came up, usually from oil officials that I would be meeting for the first time; I usually answered that the Philippine government was doing its best to address the problems of our Muslim brothers in Mindanao. My counterparts often dropped the subject afterward.

Generally, I think it helped that I managed to build personal relations with key decision makers like Saudi Arabia's Harvard-educated Oil Minister Sheikh Ahmed Zaki Yamani. He and I became such good friends that he even made me a "godfather" (in the Christian sense) to his daughter. These personal relationships certainly facilitated understanding and collaboration with the Arab countries, particularly Saudi Arabia.

My ability to develop rapport with Middle Eastern counterparts partly stemmed from my own respect for cultural differences, or at least being conscious of one's conduct as a guest in a foreign country whose customs

---

24. See, for example, W.K. Che Man, *Muslim Separatism: The Moros of Southern Philippines and the Malays of Southern Thailand* (Quezon City: Ateneo de Manila University Press, 1990).

are different from mine. For instance, on my first trip to Saudi Arabia, a middle-ranking oil ministry official invited me and two other Filipinos to dinner at his house, where he served couscous. It was my first time to encounter such a meal, and I still remember the two eyeballs of the lamb thrown at my plate by our host. He said he was giving me the honor of eating the eyeballs because he held me in his highest esteem. I took one but, since I had never eaten any eyeball in my life, I just rolled it in my mouth, not knowing what to do. After a couple of minutes, my host laughed graciously and said I could spit out the eyeball, and I did.

The benefits of the close relationship between Saudi Arabia and the Philippines went beyond obtaining crude supply for our country, since the Saudi government was also generous with advice and was even instrumental in our decision to procure oil tankers. They strongly supported the idea of a government-to-government procurement system, but they were concerned about how we would bring the crude oil to the country, and whether we had the capacity to refine it ourselves. The latter was not a problem since PNOC controlled two refining companies, but transporting the oil from the Middle East was another question.

The business of shipping oil at the time was dominated by two Hong Kong-based Chinese shipping tycoons, Y.K. Pao and C.Y. Tung.[25] Everybody knew that they controlled the oil freight business, and the Arabs did not like the prospect of having this government-to-government system "spoiled" or "diluted" by us chartering a private tanker. Given the politics in the Middle East at the time, the Arabs strongly advocated independence from multinational companies. Naturally, Saudi Arabia did not want to sell to us until we had our own means of transporting their oil. So we promised them that we would buy our own tankers, and we did. On November 20, 1974, we bought our first tanker, M/T *Diego Silang*, which had a load capacity of 760,344 barrels. Over the next two years, we bought three more tankers that plied the Middle East route; in April 1980, we bought a very large crude carrier (VLCC), the M/T *Andres Bonifacio* that

---

25. C.Y. Tung was the father of Tung Chee Hwa, the first and former chief executive of the Hong Kong Special Administrative Region.

had a capacity of over two million barrels of oil. These tankers enhanced PNOC's capability to transport nearly all of our crude oil imports from the Middle East and helped to reduce our operating costs, especially in those years when international freight costs kept increasing. It is also worth mentioning that our tankers essentially provided the training ground for many Filipino seafarers, who were eventually deployed and employed as oil tanker operators. Today, Filipinos predominate in the area of tanker operation.

Besides pressing us to use our own tankers, the Saudi government also urged us to source our supply elsewhere since there was such a huge demand for oil and Saudi Arabia—despite its status as the world's biggest oil producer—could barely cope. On the other hand, the Philippines had been totally dependent on the Middle East for all its oil imports. Hence, one of the first things we did at PNOC was to diversify our sources. We started exploring closer ties with our oil-producing neighbors in Southeast Asia.

Indonesia turned out to be most receptive toward a government-to-government relationship. Lt. Gen. Dr. H. Ibnu Sutowo, the head of PERTAMINA (Indonesia's national oil company), readily approved our proposed Petroleum Cooperation agreement that the PNOC and PERTAMINA eventually signed on February 19, 1975. Incidentally, Sutowo is credited for designing the production-sharing agreement that significantly changed the relationship between an oil-producing country and the foreign oil exploration companies. In the past, foreign companies would just walk into a country that has a potential source of oil and tell the government, "I will give you this much if you let me drill this area." Sutowo changed that for Indonesia and came up with a new system that effectively said, "Ok, I will allow you to explore this area, but the government will have a majority share. You collect all of that oil and price it accordingly, but half of the profits will go to the government and the other half we will share with you under a production-sharing agreement." Many oil-producing countries subsequently followed the Indonesian example, and our own service contract system was based on a modified production-sharing agreement. Sutowo was one of the most respected individuals in the world oil industry, although he had no background in oil. He was a medical doctor by training and served in the military during Sukarno's time. That's why they called him "General Doctor." Sutowo and I eventually developed a close personal friendship.

Our bilateral cooperation with Indonesia opened the door for exploratory talks with the other founding members of the Association of Southeast Asian Nations (ASEAN)—Malaysia, Singapore, and Thailand—on the matter of regional cooperation and mutual assistance in the development of petroleum resources. These exploratory talks culminated in the establishment of the ASEAN Council on Petroleum (ASCOPE) on October 15, 1975. We agreed that, in case of supply shortfalls, the countries with excess supply would share with the others. Although ASCOPE turned out to be largely symbolic in that no occasion necessitated the implementation of agreements on supply shortfalls, it was significant for non-oil-producing countries like Thailand, Singapore, and the Philippines, which achieved a level of comfort with the assurance that we could lean on oil-producing-ASEAN countries for assistance should a shortfall in crude supply happen again. ASCOPE's value lay in providing a vehicle for member-states to regularly discuss matters concerning oil supply. I remember how Thailand requested PNOC's assistance on how to develop an Oil Price Stabilization Fund.[26] I asked Orlando "Orly" Galang, our supply services director at PNOC, to make himself available to the ASEAN countries in the implementation of ASCOPE agreements, since Orly was very knowledgeable especially on the structuring of oil prices.

Another milestone achieved in the same year was the oil-supply agreement with the People's Republic of China, which came after the Philippines normalized ties with China in June 1975. China agreed to sell crude oil at "friendship prices." I remember, when the first shipment from China arrived at the port of the Bataan refinery, the engineers found out that Chinese crude had little gasoline content, and was composed mostly

---

26. The Marcos government, upon recommendation of the Ministry of Energy, established an Oil Price Stabilization Fund in 1984 to cope with volatile peso-dollar exchange rates. See Chapter Five of this book.

of kerosene, diesel, and fuel oil. Mobil Oil, which still had a management say in the Bataan refinery then, refused to accept the shipment! Can you imagine a tanker full of crude oil and yet wasting time at the dock because the crude was deemed unacceptable? I got so frustrated that I called up my assistant for the Bataan refinery, Jovy Jovellanos, and told him jokingly, "Bataan refinery will not accept the oil, so you better go there and drink it!" Jovy used his engineering skills to convince the American managers that with a little adjustment in the refining system and by mixing some high-grade crude oil with the Chinese crude, the new mixture would meet the special requirements of the refinery. Not long afterward, I suggested to Jovy to convince the American managers of Mobil Oil to become consultants for PNOC; after all, we had a common interest in running the Bataan refinery. The Americans accepted the offer and eventually relinquished the refinery's management to us, thereby making the Bataan Refining Corporation a wholly Filipino-managed company under PNOC.

One neighboring oil producer that we were unable to forge an agreement with was Brunei, which had not yet joined the ASEAN when we established the ASCOPE in 1975. But even a bilateral agreement with Brunei was impossible, and I attribute this primarily to the fact that Shell controlled its oil production. In fact, when the Sultan of Brunei visited Manila, I wanted to see him, but somebody told me that I had to go through Shell! Can you imagine that? A private foreign oil company was effectively screening Philippine government officials who wanted to meet with Brunei's ruler. I had no choice but to see the Shell executive who was traveling with the Sultan, and to explain to him my intentions, so to speak. Like a dutiful gatekeeper, he asked what I needed from the Sultan and, of course, I said, "Oil." Nothing came of that meeting.

Looking back, in all these tough negotiations for oil, one event helped seal the country's supply security, and this happened in December 1980. The OPEC ministers were meeting in Bali, Indonesia. On instinct, I decided to invite all of them to come over and spend the Christmas holidays in Manila afterward. To our extreme pleasure, four accepted the invitation: Dr. Mana Oteika of Abu Dhabi, Abdul Karim of Iraq, Ali Khalifa al Sabah of Kuwait, and Sheikh Yamani of Saudi Arabia. These were the heavyweights of the oil world, and the Philippines managed to bring them over. It was a diplomatic coup of sorts. The Ministry of Energy planned for the visit to the last detail. We mobilized four project teams and made sure that every aspect was ironed out—airport reception, limousines and security, escort and red-carpet treatment, meetings with high government officials and business leaders, and other perks of a state visitor. It was such a pleasure hosting the oil ministers. They were extremely gentlemanly, very conscientious of their role as guests and were never overbearing. The visit to the Philippines made such an impact on them and made them know our country better. Incidentally, President Marcos and I had no opportunity to celebrate Christmas Day with our families because Sheikh Yamani arrived on Christmas Eve and stayed until the day after Christmas.

Securing and diversifying our oil supply was not without cost. To finance related activities like the procurement of tankers, the PNOC board suggested to the president that a special fund of one centavo per liter be imposed on the retail prices of petroleum products, and that this fund be used for energy development. The president agreed and so, in April 1974, the Oil Industry Commission ruled that Php 0.01 per liter would be added to pump prices for the purpose of creating the Oil Industry Special Fund. The legal basis for the fund was contained in Republic Act 6173 (as amended), known as the "Oil Industry Commission Act," which authorized the Commission to set up a special fund for specific purposes related to the development, production and consumption of various energy resources.[27] PNOC was in charge of monitoring, receiving, and depositing the collections to a trust account.

I designated Federico E. Puno, our senior financial analyst then, to take charge of the process. The oil companies—Shell, Petrophil, Mobil, and Caltex—reported to the PNOC their monthly sales, from which they remitted Php 0.01 per liter of oil that they sold. PNOC received these remittances and put them in a trust account with the Philippine National Bank (PNB). Since the law stipulated that control over the special fund's disbursements lay with the President of the Philippines, PNOC had to

---

27. Republic Act 6173 (as amended), "Oil Industry Commission Act," April 30, 1971.

deal directly with President Marcos whenever we needed to draw on the special fund. Every time we needed to draw on the fund, we had to justify our request. If we convinced the President, then he would issue a PD or a Letter of Instruction effectively directing PNB to release a particular amount to PNOC for a specific purpose—whether for the acquisition of an oil tanker, exploration activities, and the like.

The special fund provided PNOC with financial flexibility and a steady cash flow that helped us establish our operations (especially in building our tanker fleet) and to initiate projects related to energy development. At the time, an across-the-board increase of one centavo per liter of oil products generally resulted in the recovery of about Php 100 million in one year. That was when I realized how big the oil business was! When I was in the private sector, handling millions of pesos was ordinary, but with PNOC, the magnitude of funds involved was simply unbelievable! We were talking about hundreds of millions of pesos. The taxes on oil alone already represented 20-25 percent of government income. You know, reading PNOC's financial statements was like the reading the national budget—the dimensions were so huge!

Another thing I realized when we created the special fund was that martial law facilitated the implementation of the one-centavo levy. Considering the turmoil that prevailed in the country before martial law's declaration, I do not think we would have succeeded in imposing the levy if we were still under the old political system. First, given the hardships brought about by the oil shock, we probably would have had riots, or at least daily street demonstrations and mass transport strikes. It would have been difficult to make the public understand that even if an oil company earned a billion pesos in net profit, that was in actual fact less than three centavos for every peso income. Second, it would have taken a long time before we could collect the tax since we would need to conduct public hearings and maybe even explain to Congress. Third, the oil companies would have been hesitant to collect the additional tax, given the constant rise in petroleum prices.

Besides the Oil Industry Special Fund, PNOC tried to generate income from its own operations. For instance, as we brought in more crude oil through the government-to-government procurement system, we sold some of it to the oil companies operating domestically. Especially during 1974-75, when oil prices continued to be unstable, we insisted that the foreign oil companies buy preferential crude from PNOC. The oil majors were not very happy with this requirement, but I think they acquiesced, even if grudgingly, because they also did not want to take responsibility for ensuring continuity of supply. Relying on PNOC meant that they did not have to absorb any blame in case of supply disruption. Eventually, the oil companies helped us arrange additional government-to-government supply contracts. In about a year's time, most of PNOC's crude imports were bought on a government-to-government basis.

Although this arrangement with the oil companies seemed very advantageous to PNOC, the company never really made money from the oil sales. Instead, we generated our own funds by maximizing our credit lines. Whenever we bought oil from a government, we usually negotiated for ninety days' credit. When the payment became due, instead of paying for it using our peso funds, we drew down from our dollar credit lines for 180-day availments to cover the payment and simultaneously took out a forward foreign-exchange cover from the Central Bank. With the foreign-currency risks hedged, we placed our peso funds (which would otherwise have been used for payment) in the local money market to arbitrage between the differential between the peso interest placement rates and our dollar borrowing costs. Typical peso placement rates at that time ranged from 12-18 percent while dollar interest costs were 6-8 percent. I must give credit to Edgardo "Ed" M. del Fonso, then PNOC's vice president for finance, and the rest of our finance people; they were really good. It also helped our cause that we were able to dispose the crude oil as quickly as possible. As Ed used to say, "The time to turn inventory into cash was shorter than the time needed to pay the banks." At some point, I learned that there was a joke going around the office that PNOC was not an oil company but a pseudo-bank, in the sense that Ed's group simply made our income by using oil operations as an excuse.

Another source of funds was petroleum retailing. We aggressively marketed Petron as the brand name to trust. The government's decision to make Petrophil the exclusive supplier of petroleum products to all government departments, bureaus, offices, and agencies[28] also enhanced Petron's position. We established the Petron Tires, Batteries and Accessories (Petron TBA) Corporation in January 1975, to expand the services of our gasoline stations. Through such efforts, Petron quickly captured 40 percent of the local oil market. Petron was a money-making machine, so to speak, and many of our important expenditures, including the salaries of PNOC's management team (except mine) and consultants, were paid out from Petron's coffers.

These fund-generating schemes enabled us to undertake major activities and build PNOC's capacities in the first two years of its existence. Despite our initial budget of Php 200 million, PNOC did not depend on the national government for its operations and capital expenditures. By December 1975, PNOC had established substantial capability in the major areas of the oil trade: refining and manufacturing, marketing, shipping and transport, and storage. The company had clearly established its leadership in the energy sector. This was exemplified by PNOC's designation as the lead government agency for the Energy Conservation (ENERCON) Movement that President Marcos initiated in December 1975.

But an even more crucial mission awaited us in 1976. The president, having seen what we could do, mandated PNOC to lead the way in developing the country's total-energy capabilities. Thus, from a mere national oil company, PNOC was transformed into a "Total Energy Company."

---

28. "Petrophil sole supplier for govt agencies," *Business Day*, June 6, 1974.

CHAPTER TWO
WORKING for ENERGY SELF-RELIANCE

As engineering students, my friends and I used to go on field trips and seminars to the Caliraya, Ambuklao, and Binga hydro-electric systems all over the country. My favorite was the 32-megawatt pump-storage facility at Caliraya. I would go there, take in all the view, look at the staff housing, and tell myself, "This is my dream. I want to work here and live in one of those houses." Never in my wildest imagination did I think that, one day, Caliraya would form just an inconsequential part of the comprehensive energy program that I would implement as energy minister. I don't think any other engineer in the country had the fortune of administering sophisticated engineering projects such as those that I became exposed to—the nuclear plant, geothermal development, coal-fired plants, hydro-electric systems, and the nonconventional energy schemes. I did not do the engineering myself, but I learned a good deal about these systems. I was lucky.

GZV greeting Rumanian Ambassador Filip Tomolescu at the turnover of the oil rig donated by the Rumanian government. (August 19, 1976, Barrio Salvador, Santiago, Isabela)

GZV at the dock of the Bataan Refinery. (April 1976)

Since the beginning of the oil crisis, President Marcos was concerned about reducing the country's dependence on imported oil. The oil shock of '73 exposed the vulnerabilities of Third World countries like the Philippines, whose economic development rested on the steady supply of affordable oil. Everything we did—running our industries, producing commodities, transporting goods and services, not to mention enjoying the modern comforts of living—was powered by oil. But the crisis made us realize that the world's oil supply had become so volatile that dependence on oil imports was no longer a reliable option to fuel the country. If we were to make progress, we had to develop our own energy sources.

With this goal in mind, President Marcos made two important moves aimed at enhancing governmental capacity to undertake energy-related activities. In March 1976 he issued Presidential Decree (PD) 910, creating the Energy Development Board (EDB), to integrate in one body all policy formulation and regulatory functions relating to energy exploration and development.[29] The EDB absorbed the powers and functions of the Department of Natural Resources and the Bureau of Mines in relation to the administration, exploration and development of coal, geothermal energy, natural gas and methane gas. The president appointed me as concurrent chairman of the EDB, with Cabinet rank. A month later, in April 1976, he converted the Philippine National Oil Company (PNOC) into an energy company by virtue of PD 927, which amended the PNOC Charter and expanded the company's responsibilities to include the acceleration of exploitation, exploration, and development of indigenous energy sources, especially non-oil resources.[30]

---

29. Presidential Decree 910, "Creating an Energy Development Board, Defining Its Powers and Functions, Providing Funds Therefor, And For Other Purposes," March 22, 1976.
30. Presidential Decree 927, "Further Amending the Charter of the Philippine National Oil Company (PNOC) As Provided For In Presidential Decree 334, As Amended, And For Other Purposes," April 30, 1976.

These developments laid the basis for intensified exploration and development of the country's indigenous resources that had been overlooked and neglected over the past decades. A good example is coal. The presence of coal deposits in many parts of the country such as Cebu, Batan Island off Albay, Malangas in Zamboanga del Sur, and Semirara Island off Antique, had long been known, but there were no reliable estimates of reserves available. Exploitation had also been minimal and confined mainly to Cebu and the Bicol provinces. On the other hand, the existence of geothermal energy resources had been equally evident in many places, but not until the onset of the energy crisis did efforts at serious development begin.

We can say that in 1976 the strategy of PNOC shifted from an oil-procurement mission to an energy-development mission, encompassing the entire field of energy and the initiation of a broad range of energy projects. Besides its original mandate of providing adequate and stable supply of crude oil and petroleum products to the country, the PNOC was now expected to lead national efforts at developing indigenous energy resources. Thus, the exploration and development of the country's petroleum, coal, and geothermal resources, along with energy conservation and petrochemical development, constituted PNOC's new thrusts.

To me, aspiring to energy self-sufficiency was a natural stage in the country's development. After all, in the early Seventies, the National Power Corporation (Napocor or NPC) had already engaged in limited geothermal experimentation. For our part, after two years of operations, I had become very confident of PNOC's capacity to take on this new challenge. However, before we could make any move, we had to address the major gap in knowledge about the extent of the country's resources. As I mentioned earlier, there was an appalling lack of consolidated data on different energy sources, and so we commissioned scientific and technical studies to aid the creation of a systematic inventory of indigenous energy sources. We had to begin from scratch, and since we wanted to reduce our dependence on imported crude oil by at least 50 percent especially in the power sector, we looked into developing alternative sources such as hydro, geothermal, coal, nuclear, and others.

Of course, there was some semblance of data collection on coal resources through the Bureau of Mines. Before obtaining a concession, the mining company had to file an application containing information about the estimated amount of available coal deposits in the mining prospect and the expected volume to be extracted by the company; the Bureau of Mines issued a permit based on such information. So we had a fairly good knowledge of the amount of coal, but not of its quality. There are various types of coal. Being a geologically young country, we were told that we only have young coal of the lignite class, not the old coal—the bituminous and semi-bituminous type. Perhaps this is also one of the reasons why we do not have extensive oil and gas deposits, because we are still a geologically young country.

In the case of hydro resources, Napocor had available hydrologic data from various areas of the country. Although the data was not sufficiently accurate to be used as basis for designing the required power plant, nevertheless they gave you an idea of what the possibilities were in these specific areas.

We were completely lost, however, in the area of geothermal energy. We knew we had geothermal sources, but since we were only starting to realize that we had them, we simply had no knowledge of the extent of these resources. If not for the need to develop an energy policy, we would not have seriously studied geothermal's potential as an alternative energy source.

The same was true for nonconventional sources. Take bio-gas: no matter how small its contribution to the national effort, we had no knowledge about it. We knew that animal waste was a good source but there was no inventory, and not even an attempt to explore its potential. Another source we stumbled upon was marsh gas, a somewhat low-grade gas that comes in relatively small quantities. It can be used for cooking, and although it can be found in the hinterlands, people can use marsh gas anywhere.

We also realized the distinct qualities of each of these resources. For example, geothermal energy is site-specific: you have to use that energy in the site. On the other hand, coal or oil can be transported. Hydro is another site-specific energy, although its finished product—electricity—can be transported.

In short, before we could start the energy plan, we had to go into the consideration of all the basics because we did not have available data. It took two to three years before we developed our knowledge, and we did not even have trained people for this. We had to look to other institutions whose engineers, chemists, and scientists dared to delve into the new areas of energy. That is why it was not possible to come out with an energy policy immediately. All available data on these energy sources had to be put together piece-meal.[31]

Personally, since I valued comprehension—of the problem, of the job—as a crucial element in management, I put importance in data-gathering and scientific studies. I also liked working with people who were flexible, and who had the willingness to learn and the readiness to adapt to changes. I applied this perspective to PNOC as we went about fulfilling our new mandate.

One of the first things we did was to look into our own human-resource base. Many PNOC executives were aghast when told that they would have to look beyond the oil business! You see, most of our people were knowledgeable in this field, but their knowledge of non-oil resources was practically nil. Thus, I decided that the company should make a conscious effort to tap the country's top geologists, engineers, physicists, and other scientists to help us carry out our mission. Although PNOC was already working with foreign consultants for various types of technical assistance, I made it a matter of policy to organize our own pool of Filipino scientists and engineers who possessed sufficient expertise, or at least the aptitude to learn new concepts from their foreign counterparts. Technology transfer was an important element of PNOC's relations with foreign consultants, and we did not hesitate to send our people abroad for training.

On the organizational side, we created two new subsidiaries that would enable PNOC to carry out its added mandate: the PNOC Exploration

---

31. In this regard, I want to acknowledge the contribution of Gary Makasiar in accumulating and synthesizing the data on energy resources and coming up with the Philippine Energy Plan.

Corporation and the PNOC Energy Development Corporation. Over the next four years, we set up more subsidiary companies as we explored alternative energy sources simultaneously. I applied what you call the "conglomeration approach" to management, purposely organizing separate entities so that each could focus on the job required. By doing this, both PNOC (as the central management unit) and its various subsidiaries remained lean and focused in carrying out their respective assignments.

It was also in this phase of our development that I realized how fortunate I was that PNOC had a very competent and professional staff. All I had to do was know how to tap them and give them the right tasks that would bring out the best in them. Having been a CEO of different companies for more than twenty years, I had learned to rely partly on my instinct in the choice of people for particular assignments. I also had a keen sense about people not being able to get together or which teams could work well. I looked to my instincts in probing this during staff meetings, in executive sessions, the way I was getting answers or not getting solutions.

Whenever people asked about my management style, I always described myself as an uneducated manager, since I never had formal management training. An essential aspect of my management principle, though, is the consideration of Filipino culture, especially what we call in local parlance as *pakikisama*. By this I mean paying attention to people, their aspirations and capabilities, their cultural background, the informal incentives of competitive pay and rewards, including their job and family security. I am proud to say that in all my years as chief executive of private companies, whether before or after my career in government, I did not experience a single labor strike. I consider it a noteworthy achievement that, in 1991, the Kilusang Mayo Uno (KMU), under the leadership of Crispin Beltran, gave me an award of recognition (the only one ever given by KMU to an employer) for exemplary treatment of Republic Glass employees. Ultimately, management is about managing people, and the role of the manager is to steer the company to meet objectives and goals, and tapping manpower resources to achieve these. This is true of corporations and true of nations. This is what leadership is all about.

How I translate management principles into practice could be tough for my staff, however. The executives of PNOC had a lot of adjustment to do with regard to my work style. First of all, I am a one-page memo man. If they sent me long memos, I gave it back to them saying, "Digest its essence in one page." And it better be good and substantive. I know I had people tearing their hair out trying to put all they wanted to say in one-page memos. But my attitude was, either you had something solid to say and knew how to say it, or you had nothing much to say.

When I was dissatisfied with the memo, I would then ask for formal presentations, which meant the works—computer printouts, slides, acetates, and charts. My colleagues' training in the oil majors helped them make these presentations very slick at times. I would preside over these presentations, questioning numbers and always looking for bottom lines. I developed my own set of tools for decision making, for probing, for sifting through management jargon and financial figures. Since I am an engineer, I was comfortable with figures if they made sense, and the only figure that made sense to me was the bottom line. So whenever I read their reports, I often began with the last page, interpreting the numbers from bottom up.

On the other hand, if the memos they sent were clear, comprehensive and the decision required did not necessitate authorization from the president, my executive staff could expect to receive my response within the day or the next. In this way, I avoided backlog and managed to keep my desk clean at the end of the day.

Another cardinal rule I established among my executive staff was, nobody walked out of my room without a decision. This ensured that things kept moving. If a particular endeavor did not go right, I simply took it to mean that the basic assumptions we used in making a prior decision had changed and, therefore, the environment, the circumstances, the inputs and processes changed. The appropriate response, then, was to rework the program or project, redesign if necessary, and take it from there. It was simply a new set of givens. We were not going to lose sleep over something we worked so hard over, breaking our backs so to speak, and cry like babies because it did not work out. No. We were simply going to start a new program or project.

I also required that whenever I was in the office, everyone else had to be around. Everyone. The only exception I allowed was if somebody had a prearranged meeting or emergency elsewhere. I ran my executives like a commando squad, if you will, because we had all the necessary communication facilities. Each car was outfitted with special walkie-talkies, telephones, and transmission systems, so that I could get anyone I wanted on the line anytime I wanted, whether during or after office hours, Sundays and holidays. My colleagues got used to my calling any time of the day or even during ungodly hours, wives notwithstanding.

My meetings were short and to the point. I never indulged in small talk. Sure, in socials or parties, I could be very charming, but not in business meetings or executive briefings. I always told them, "If we have nothing to talk about, let's quit the meeting." I also was not the type who would backslap the executive and say, "How's the wife?" I kept the personal out of business. In that sense, I appeared unfriendly, aloof, or even an SOB to some of my men, but it did not matter to me.

Luckily, my instincts served me well at PNOC, and what made me more proud of my colleagues was that they showed they could handle new challenges with ingenuity, gusto, and dedication.

Oil Exploration

In the past, petroleum exploration in the Philippines was done in trickles and without success. Ever since Smith Bell bore the first hole in 1896, some 301 shallow wells had been drilled in the country. In the Fifties and early Sixties, private companies operating under the concession system drilled eighty percent of these wells, all of which turned out to be either dry or not viable for commercial development. Although the country gained in terms of information, since the Philippine Bureau of Mines gathered considerable data from these failed drilling operations, we lost in terms of potential investment as the multinationals pulled out, leaving the country starved of capital and technology. For several years afterward, drilling was at a standstill.

Despite such failures, Filipinos persistently believed that, since the Philippines lies along the so-called oil path of Indonesia, Malaysia, and even Thailand, we have some oil lying around somewhere in this archipelago. In fact, Sabah's potential as a rich source of oil partly drove the Philippines' claim over this territory, which severely strained Philippine-Malaysian ties for many years.

But it was an external event that sparked a renewed interest in oil exploration. The Arab-Israeli war of 1967 led to the temporary closure of the Suez Canal, an important trading route plied by tankers that carried oil from the Middle East to the rest of the world. With crude and freight rates rising, the Philippine government auctioned off fifteen concession blocks to about sixty Filipino exploration companies in 1969. Unfortunately, drilling operations remained sporadic due to lack of high-risk capital and expertise, as well as the absence of government incentives. Until the early Seventies, in fact, the government had shown little interest in oil exploration. The Petroleum Act of 1949 (Republic Act 387), an antiquated law that wrongly presumed that such a capital-intensive, high-risk and speculative venture could be undertaken successfully by a single individual or company, also stifled oil exploration and only resulted in the financial exhaustion of petroleum concessionaires. Moreover, the old concession system practically closed the door on interested foreign investors. Hence, all exploration activities were financed primarily by local capital, which was stretched to its limits. Moreover, the joke around town then was that, oil exploration was simply a "stock-market operation"—discovery was dependent on the volume of speculative trading.

This situation changed with President Marcos's issuance of PD 87 or the "Oil Exploration and Development Act of 1972." This landmark decree not only instituted the service contract system, but also created the Petroleum Board as the administrative body that devised the incentives to attract local and foreign exploration firms. These incentives were: exemption from tariff duties and compensating taxes on importation of machinery, spare parts and materials for operations; repatriation of capital investment actually brought into the country; and retention or remittance abroad of foreign-exchange earnings in excess of operating requirements.[32]

---

32. Presidential Decree 87, "Oil Exploration and Development Act of 1972," December 31, 1972.

The success of PD 87 in creating a favorable investment climate became evident with the entry of foreign and local oil exploration firms such as Caltex (Chevron-Texaco), Philips Petroleum, Superior (Arco), PODCO, Champlin, Amoco-Mosbacher, Husky-Cities Service, Pioneer Bay, and others. In April 1974, I reported to the president that, "The sum of commitments entered into under the present system has reached an amount of US$ 72 million or Php 504 million in slightly more than a year's time, surpassing the amount of Php 147 million spent under RA 387 by local companies in the past twenty-four years."[33]

PNOC's mandate of promoting the exploration, exploitation and development of local and petroleum sources was strengthened when President Marcos issued PD 469 (amending PD 87)[34] in May 1974, designating me as chairman of the Petroleum Board, the interagency body authorized to enter into service contract agreements with exploration companies. Now, just like when I came into PNOC, I was fortunate to work with very good people who helped to set up PNOC's exploration arm.

The first was the late Wenceslao "Pat" de la Paz. He was a tax lawyer for Esso (Philippines) but I made him director of the Petroleum Board, which implemented the service contract system. A postgraduate-degree holder from New York University, Pat was a very good lawyer; he drafted the presidential decrees that we needed, although these drafts had to first go through Solicitor General Estelito Mendoza before I brought them to the president for signing. Later, I assigned Pat to head the Bureau of Energy Development (the Petroleum Board's successor) under the Ministry of Energy. He was a most efficient and effective assistant who was instrumental in developing our acquaintance and excellent relationship with the foreign companies that entered into oil-service contracts with the government. In the later years of the Aquino administration, Pat was designated head of energy affairs, becoming a de facto "energy czar" and filling the void that resulted from the abolition of the Ministry of Energy.

---

33. Geronimo Z. Velasco, "Memorandum to President Ferdinand E. Marcos," April 24, 1975.
34. Presidential Decree 469, "Amending Presidential Decree 87, Otherwise Known As "The Oil Exploration and Development Act of 1972," May 23, 1974.

The other person who contributed substantially to our attempt at actual exploration was the geologist Arthur "Art" Saldivar-Sali. It was by accident that I came to know him. One day, I was reading the newspaper when I saw this item about his having returned to the country after four years of PhD studies at the University of London's Imperial College of Science and Technology. The article said that Dr. Sali was back at the University of the Philippines, teaching at the College of Engineering. I thought to myself, "We need to get this guy! We have to have a geologist!" You see, there was no geologist in the old Esso and Filoil companies. Can you imagine an energy company that did not even have a geologist? I immediately ordered my staff to check out Dr. Sali.

When I met Art at the PNOC office, the first thing I asked him was whether he knew how to look for petroleum. He then explained that he had studied petroleum geology as an undergraduate at UP, but his post-graduate degree from Imperial College was in engineering geology which, he said, was about "developing the necessary infrastructure once one finds petroleum, gold and other kinds of mines, and similar resources." I immediately said, "You're the person we need! Definitely, we will search for oil, but what do we do afterward? Why don't you join us at PNOC?"

But Art said he owed UP eight years of service in return for his studies abroad. Right away I called up then-UP President Salvador P. Lopez, but he did not agree to release Art because it would set a bad precedent for the UP faculty. So the next thing I did was to have my secretary call up Malacañang. I talked to President Marcos, explained my predicament and secured Art's release from his UP obligations without him having to pay a single centavo, since he was being seconded to another government organization anyway. That was how determined I was to get him.

Besides being the deputy director of the Petroleum Board, Art Sali played a crucial role in setting up PNOC's exploration team. By then, efforts were already under way to enhance our capability to undertake oil exploration. We were lucky because Indonesia agreed to help us out. In October 1974, PNOC and PERTAMINA signed a technical assistance agreement in which the latter would provide consulting and training services to PNOC in the matter of oil exploration. I took advantage of this to send Art, Pat de la Paz, and another geologist to Indonesia for about two months, traveling to the oil fields from Kalimantan to Irian Jaya to learn the ropes of exploration. Subsequently, we set up the exploration department, and initiated preliminary negotiations with Romania for the purchase of an oil-drilling rig. Then a team of exploration advisers from West Germany arrived for a two-year program on basins evaluation and exploration work. PNOC availed itself of the advisers' services through the 1971 Technical Cooperation Agreement between the Philippines and the Federal Republic of Germany.

Most foreign exploration firms operating under the service contract arrangement at the time were offshore. Now, offshore exploration is very expensive, and I felt that the capital and technology requirements were simply beyond the government's capability. So we decided to concentrate on onshore exploration, and the first area we identified was Cagayan Valley, a large basin that was hardly touched by oil majors like Esso, Caltex, and Mobil when they undertook exploration activities in the Fifties.

In April 1975, PNOC became the first entity to undertake surveys of the area's petroleum potentials, mobilizing more than 650 geologists, gravimetric and seismic teams to scan the Valley. For the seismic and gravity surveys, PNOC contracted the services of Delta Exploration Company. A 300-man PNOC team set up camp in Ballesteros, Isabela, to undertake geophysical and geological surveys in the Cagayan Valley. When we acquired a deep-drilling oil rig from Romania in the same month, that was it. There was no stopping us from trying our hand at oil exploration.

By 1976, we had acquired sufficient confidence to spin off PNOC's Exploration Department and set up the PNOC Exploration Corporation (PNOC-EC). With Art Sali as head, the company concentrated on onshore oil exploration to complement the private sector's offshore exploration. Of PNOC-EC's initial capitalization of Php 300 million, Php 253 million came from the Oil Industry Special Fund. But I made sure that unlike PNOC, PNOC-EC would not be a wholly government corporation; about 5.6 billion shares were subscribed to by 5,028 private corporations and individuals, representing 18.7 percent of the new company's authorized capital stock.

As PNOC-EC, we acquired three service contracts in Cagayan Valley with a total area of 780,000 hectares and a budget of Php 35 million. Then, we mapped out a continuing program of drilling twenty wells over five years. This was a whole new ballgame for Filipinos and we were very excited about it. Our first well was a success, as we discovered gas-bearing zones. Buoyed by our discovery, we drilled five more wells except we did not find anything more. At any rate, the National Power Corporation developed our first well and today it is producing gas that partially supplies the electrification of Cagayan Valley.

PNOC-EC's activities encouraged other prospectors such that, by 1980, seventy-seven companies had been awarded operating contracts to authorize their participation in various aspects of oil exploration (see Table 2.1: Performance Indicators in the Oil and Gas Sector). All these efforts led to the discovery of eleven oil-bearing structures, and the development of the country's first commercial oil field offshore, the Nido Reef Complex.[35]

The country found cause to celebrate when oil was discovered in Nido, Palawan on March 9, 1976; the story behind Nido, however, actually began much earlier. I think it was around late November or early December in 1975 when I received an overseas phone call from an aggressive and persistent American named Allen Hatley. I did not know him from Adam; he was in New York, but he kept calling up that I finally had to accommodate him. He said he was from Cities Service, a medium-sized oil company, and he wanted to enter into a service contract agreement for offshore exploration in Palawan. Hatley insisted that we sign the contract before December 15, 1975, or else his company would remove his budget for drilling oil in the Philippines.

I did not need much convincing. I said, "Why don't you come over so we can discuss it?" He asked, "When can I come over?" I replied, "Anytime you want." He arrived in a week's time. We discussed his proposal and I probed into the contract details. At the end of our discussion, I assured him, "You have a contract." After the American left, I instructed Pat de la Paz, my assistant at the Petroleum Board, to prepare a memorandum for

---

35. *The National Energy Program, 1981-86* (Makati: Ministry of Energy), p. 17.

Table 2.1
Performance Indicators in the Oil and Gas Sector

Year	Number of wells		Footage (000FT)	Seismic lines (KM)	Number of service contracts		
	Onshore	Offshore			Geophysical permits	Geophysical survey contracts	Service contracts
1973	—	2	20,996	11,744	—	—	8
1974	1	2	25,320	16,020	—	—	3
1975	—	10	88,011	8,832	—	—	2
1976	3	6	81,882	5,288	1	1	9
1977	6	9	127,332	10,064	1	3	2
1978	5	7	89,012	22,164	2	7	3
1979	6	17	172,252	15,403	5	8	2
1980	7	14	173,455	20,930	6	10	4
Total	28	67	778,260	110,445	15	29	33
Average (1973-1976)	1	5	54,052	10,471	1*	1*	6
Average (1977-1980)	6	12	140,513	17,140	4	7	3

*Started 1976

Source: The National Energy Program, 1981-86. Makati: Ministry of Energy. Republic of the Philippines

President Marcos requesting authorization for me to sign an agreement with Cities Service. You see, only the president had the authority to grant such concessions.

That same evening, I brought the memorandum to the president, which he signed right away. The following day I saw Hatley. I said, "Come on, let's sign the contract." Was he surprised! Imagine, he had a contract twenty-four hours after our first meeting. Securing a government contract in twenty-four hours was unheard of in the oil industry. We could not have done that were it not for two things: first, the president's complete trust in me and, second, martial law. You can say that we were instant decision makers. Prospective investors had to present a viable proposal that I would clear with the president so we could implement a contract. You know, that was a concession involving a natural resource of the country; only the president could give you the authority to utilize that. And since he was the sole decision maker for the country, I did not have to worry about explaining myself to anyone else.

The service contract was officially signed on December 17, 1975. Hatley himself marveled at the speed with which we wrapped up the deal. As he wrote almost twenty years later, "In retrospect, it was a truly remarkable accomplishment that could only have happened after martial law and the organization of the Philippine government's Petroleum Board, and its successor, the Bureau of Energy Development."[36]

In early 1976 Cities Service, through Philippine-Cities Service and Husky (Philippines) Oil Inc., began drilling in the wide-open sea around Palawan and, on March 9, they struck oil at the Nido-1 well. The stock market soared, the whole economy was in euphoria, and I was grinning from ear to ear. Nido-1 opened a new phase in oil and hydrocarbon exploration in the Philippines. The Palawan oil find reaffirmed the underlying policy that had been articulated right at PNOC's birth—the need to reduce our dependence on imported oil. Being the first major oil find in the Philippines, the Nido structure initially left us at PNOC giddy

---

36. Allen G. Hatley, "The Philippines: Finding Oil Where It Shouldn't Be," in *The Oil Finders*, Allen G. Hatley, Jr. (Ed.) (Tulsa: American Association of Petroleum Geologists, 1992), p. 64.

with euphoria, especially when commercial production began in January 1979.

It was good timing too, since the world was in for a second oil shock. Several months later, the second oil crisis came following the fall of the Shah of Iran and the Iran-Iraq war. Nido-1's output of 40,000 barrels a day helped cover the shortfall that we encountered due to rising oil prices. Unfortunately, Nido-1's initial output of 40,000 barrels a day could not be sustained for more than a year of operations. As was explained later by Art Sali, the oil structure was too small for long-term commercial production, and its potential was such that the well was good for only one year.

Despite this seeming setback, total production from Nido reached 12.2 million barrels of oil by the end of 1980, which generated for the country US$ 414 million in foreign exchange savings. In addition, domestic crude-oil production displaced ten percent of our crude imports in 1979 and five percent in 1980.[37] More important, other discoveries followed such as in Cadlao, Matinloc, Tara, and Libro (all in Palawan). As Art Sali once explained to me, "These discoveries were giving more clues to the geology of the area, and they led eventually to the discovery of Malampaya, which is the biggest gas find so far." As PNOC-EC got deep into oil exploration, we accelerated our search for coal, geothermal, and other non-oil resources. I am also proud to say that, because of his work for PNOC-EC, the Petroleum Board, and the Bureau of Energy Development at the Department of Energy, Arthur Saldivar-Sali was named one of the Ten Outstanding Young Men (TOYM) of 1978.

By then, too, the country's energy development program had expanded so rapidly and had become vast enough to necessitate the creation of a single agency that served as the primary instrumentality that would formulate and implement the program on a coordinated basis. Thus, on October 6, 1977, President Marcos issued PD 1206,[38] establishing the Department of Energy and designating me as its secretary. The decree also

---

37. *The National Energy Program*, p. 17.
38. Presidential Decree 1206, "Creating the Department of Energy," October 6, 1977.

placed under my jurisdiction the Bureau of Energy Development (BED), which administered the national program for the exploration and development of various forms of energy sources, and the Bureau of Energy Utilization (BEU), which took over the functions of the abolished Oil Industry Commission in regulating petroleum and other energy-related business activities. I assigned Pat de la Paz to head the BED, and Orlando Galang to the BEU.

Geothermal Energy

Of the alternative-energy programs that the government embarked on in response to the oil crisis, geothermal development yielded the most substantial results in the shortest time possible. Geologically speaking, the Philippines lies on a high heat-flow region, or the so-called Circumpacific Volcanic Belt, which is a geologic advantage.[39] A power-generating system based on geothermal steam can displace conventional oil- and coal-fired power plants, a 110-megawatt output of geothermal power saves the approximate equivalent of one million barrels of fuel oil and, unlike fossil fuels, geothermal energy is theoretically inexhaustible.[40]

The country's potential for geothermal power, and specifically the use of geothermal energy for generating electricity, was demonstrated on April 12, 1967, when the first light bulb was powered by geothermal steam in an experimental laboratory in Tiwi, Albay. Commercial geothermal development got a boost in the same year, with the passage of Republic Act 5092 ("Geothermal Energy, Natural Gas, and Methane Gas Law") that provided access to prospective areas as it facilitated the granting of geothermal exploration permits and leases. By 1969, geothermal steam had been successfully discharged in Tiwi and utilized to power a 2.5-kilowatt noncondensing geothermal plant. This paved the way for the government's decision to start the commercial exploitation of the resource.

---

39. "Geothermal for Power," *Energy Forum* (Makati: Philippine National Oil Company), Volume One, 1985, 11.
40. Philippine National Oil Company, The 7th Year: Towards Energy Self-Reliance (1980 Annual Report).

Subsequently, President Marcos issued presidential proclamations that set aside known geothermal areas as national reservations.

My first encounter with the business of geothermal exploration occurred sometime in 1970, when I was in Honolulu on a business and vacation trip with my family. One day, I received an overseas call from my good friend Alejandro "Alex" Melchor who, besides being President Marcos's executive secretary, sat on the board of directors of the National Power Corporation. Alex asked me to establish contact with the Union Oil Company of California (Unocal) in the hope that the company, known for geothermal development in California's Sonoma County, might be interested in tapping the Philippines' geothermal potential. Using my Castle and Cooke connections, I established contact with Mr. Fred Hartley, then chairman of Unocal.

When Hartley heard that the Philippine government wanted to meet with him, he said, "Oh, I would like to talk to them. Those guys have a fantastic geothermal potential, but they're not doing anything about it. They're just sitting on it!" I called up Alex at once and relayed the message; he got angry and said, "Ok, we will go there next week." He arrived in Honolulu and dragged me to San Francisco to visit Union Oil's geothermal installations in Santa Rosa. I did not even know Fred Hartley. Nor did I think that one day I would be deeply involved in geothermal activities.

The timing of the meeting could not have been more propitious, as it took place on the eve of the first oil shock. On September 10, 1971, Unocal secured a service contract from National Power for the exploration and development of geothermal energy through a local subsidiary, Philippine Geothermal Inc. (PGI). I tell you, that contract really favored Unocal! At the time, there was no clear basis for pricing the steam other than Union Oil's demand that we pay them for their technology in extracting steam. Eventually, both parties agreed to base the price of steam on current oil prices. But when the oil shock came, the price of geothermal steam also rose disproportionately such that, to my mind, PGI's windfall profits were too much compared to the company's investment risks. In 1974, we started renegotiating with Union Oil and PGI to revise the contract with Napocor. As a result, the escalation formula was changed from one tied to oil to a formula based on the movement of the US wholesale and consumer price indices for the dollar costs, and the Philippine consumer price index for the pesos costs, resulting in a more moderate escalation clause. I would have wanted even better terms, but I also recognized that it was the maximum concession that Union Oil was willing to make, considering that the contract had already been sealed. Had the company insisted on the original contract, we would have lost more. At any rate, the result was the geothermal fields of Tiwi (Albay) and Makiling-Banahaw (Laguna). The initial geothermal production of 110 megawatts in Tiwi in 1976 and Unocal's interest in tapping the Makiling-Banahaw area made us all optimistic about going full blast with PNOC's own geothermal development activities.

In 1974, the New Zealand government awarded a NZ$ 15-million geothermal exploration technical assistance to the Philippines. Chosen by the New Zealand government as aid implementing agency was the Kingston Reynolds Thom Allardis (KRTA), which had extensive experience in geothermal exploration and development. Through this aid, PNOC explored two geothermal fields in Tongonan (Leyte) and Palinpinon (Negros Oriental), which turned out to have enough reserves to support commercial-scale power-generating plants. It was also under this program that the New Zealand government donated a drilling rig to the Philippines.

After the joint New Zealand-Philippine team confirmed that another 110 megawatt-plant could be set up in Tongonan, the PNOC Geothermal Group established the steam gathering system to enable NPC to install its power-generating facilities. The Palinpinon development, on the other hand, was the first to be undertaken solely by PNOC. To this end, we set up the PNOC Energy Development Corporation (PNOC-EDC) on March 5, 1976, to undertake the exploitation of coal, geothermal energy, uranium, and other non-oil resources. It had an initial capitalization of Php 300 million. The man I assigned to head the company was Pablo "Pabling" V. Malixi.

Pabling was the vice president of Esso (Philippines), the highest position that a Filipino ever occupied in that company. During those initial briefing sessions I had with the executives of the old Esso (Philippines),

Pabling and I did not really get along well. Let us say we both had strong personalities, we were both sure of ourselves, but only one of us could be boss. For quite some time, there was tension in our relationship. But I also saw that he was an action-oriented guy, a doer, and so I decided to make him PNOC vice president, handling Petrophil's oil unit. In 1976, when we set up PNOC-EDC, I made him concurrent coordinator for energy development.

When I talked to him about becoming the point man for geothermal development, Pabling hesitated initially since he did not know anything about non-oil energy sources. He was so concerned about where to get the people, the money, and other things needed to operate the company. My reaction was, "*Ay naku, Pabling, binibigyan na nga kita ng importanteng trabaho, ang dami mo pang tinatanong!* (My goodness, Pabling, I'm already giving you an important job and you still ask too many questions!)" But I am really glad I took the risk in assigning him to the geothermal project because he did a terrific job. He was a very competitive person, and I suspect that he made use of the opportunity to show what he was capable of.

You will probably ask, what gave us the confidence to go quickly into geothermal development? There had been many scientific studies undertaken by the old Commission on Volcanology since the early Sixties, but nobody made use of them as a guide for commercial exploration. When we decided to take up geothermal development, I had the old studies reviewed. I was aware of some of these because I personally knew Dr. Arturo "Art" P. Alcaraz, former head of the Commission on Volcanology, one of the world's acknowledged pioneers of modern geothermal utilization, and "father of Philippine geothermal energy."

Dr. Alcaraz initiated the early studies on Tiwi's geothermal potential, having undertaken geoscientific investigation of the area. I met him in April 1967, while I was still with Dole (Philippines). One day, Alex Melchor (who was already President Marcos's executive secretary) asked me to go with him to Tiwi, Albay, because he wanted to show me an experimental laboratory for geothermal energy. Alex seemed very excited, so I went with him to Tiwi. There I met Dr. Alcaraz, who showed off his "toy": using geothermal steam to drive a three-kilowatt generator, he lit an ordinary light bulb. That was my first exposure both to geothermal energy and to Art Alcaraz.

Later, I found out that Art Alcaraz was more closely related to me than I thought. It turned out that he was my wife's cousin, and he became a geologist because of my father-in-law, Leopoldo Faustino, who was the first Filipino geologist and the first Filipino to obtain a PhD in geology from Stanford University. Art was an admiring nephew, so he decided to follow my father-in-law's example and became a geologist too. He graduated from the University of Wisconsin and did additional studies at the University of California in Berkeley and at Kyushu University, specializing in volcanology, seismology, geology, and geothermy. I took in Art as a consultant for PNOC-EDC and subsequently for the National Power Corporation, when I became its chairman ex officio in 1977.

By that time, I had made it a matter of policy to literally grab every scientist or technocrat available or working somewhere, and to challenge him with a problem of national dimension. Art Alcaraz helped to identify possible recruits, and one such person he recommended was Nazario "Nazi" C. Vasquez, who began his career at the Commission on Volcanology and worked alongside Art in exploring potential geothermal fields. Nazi was the only other authority on geothermal besides Art Alcaraz, and since he was working by then for Union Oil's Philippine Geothermal Inc. (PGI), we felt that his experience in field operations would help PNOC-EDC immensely. But it was not easy to make Nazi leave PGI; after all, he already had a comfortable job. We had to appeal to his sense of patriotism, saying that building the country's geothermal capability was a national challenge and we needed his expertise. Finally, he joined us in 1976 and we made him project manager for the Tongonan field. This policy of recruiting people with specialized expertise enabled PNOC to pursue and accelerate its comprehensive energy development program.

In geothermal development, the strategy was to exploit the geothermal resource in the shortest possible time. The economic value of time was given much weight in the decision-making processes. This aggressive stance meant putting in a little more risk capital than would have been needed by an orthodox and conservative approach. PNOC's strategy, therefore, was aimed at avoiding delay by eliminating time-consuming activities and accelerating critical operations, particularly the drilling of production wells. A corollary strategy was the furtherance of scientific knowledge and expertise, the ultimate objective of which was the development of a sufficient local manpower base. In this connection, we established a system of on-the-job training, complemented by exposure of Filipinos to specialized activities abroad. Pabling Malixi himself went on six-month learning missions to the geothermal fields of California and New Zealand to prepare for his job as coordinator for energy development.

In 1977, one year after PNOC-EDC was formed, the Philippines operated its first commercial-scale three-megawatt power plant in Tongonan, Leyte. Other installations followed suit, such as the eight 55-megawatt plants in Tiwi (Albay) and Makiling-Banahaw (Laguna), and two 1.5-megawatt units in the Palinpinon-Dauin area near Dumaguete City; we developed the Palinpinon plant with the help of a soft loan from the Japanese government. In the next three years, PNOC geared its program toward developing 1,331 megawatts of geothermal power in the provinces of Leyte, Southern Negros, Albay, Benguet, and Davao. The Ministry of Energy also obtained technical assistance from the Italian government to undertake a detailed inventory of the country's geothermal resources. Conducted jointly by the Bureau of Energy Development and the Italian company Electro-Consult, this resource inventory led to the prioritization of the following areas for development: Batong-Buhay (Kalinga-Apayao); Mainit Bontoc (Mountain Province); Buguias, Daklan, Acupan-Itogon, and Apo Kidapawan (Benguet); Pinatubo (Zambales); Cagua (Cagayan Valley); Montelago (Oriental Mindoro); Biliran Island (Northern Leyte); Anahawan (Southern Leyte); Mainit-Placer (Surigao del Norte); and, Balatukan (Misamis Oriental).

By then, PNOC-EDC had an expanding base of local experts working closely with foreign consultants. As we developed the different alternative-energy programs, PNOC started to attract the brightest minds in geology, geophysics, engineering, and other related fields. I am proud of the fact that in those years, practically all the honor graduates and top professors of the UP College of Engineering worked at PNOC, not to mention the geologists, physicists, and others whose expertise tremendously boosted our programs. To consolidate our efforts at building in-house expertise, in 1981 PNOC formally entered into a joint venture with KRTA, the Geothermal Technology Corporation. The contract signing was witnessed by President Marcos and New Zealand Prime Minister Robert Muldoon.

This joint venture enabled us to enjoy more fully the benefits of technology transfer from our New Zealand counterparts since, under the setup, key people from KRTA came to the Philippines so that a greater number of PNOC-EDC personnel got the opportunity to work with them. And since working with expatriates in actual field and office operations had been one of the most effective means of transferring technology, PNOC-EDC personnel acquired varying degrees of self-reliance in the different stages of geothermal exploration and development.

You must be wondering why New Zealand played such a big role in our geothermal program. Well, as Arturo Alcaraz once explained, our geothermal steam is similar to New Zealand's, which is hot water-dominated. On the other hand, the geothermal steam of Italy (the world's first user of geothermal) and the United States (the world's current biggest user) are vapor-dominated. As such, New Zealand is considered as the pioneer in hot water-dominated geothermal steam technology.

Geothermal development is heavy on front-end expense because you have to drill wells, pipe them, and build the power plants nearby. However, the advantage of a geothermal plant over a coal- or oil-fired plant is, after you have set up the infrastructure to generate power, the steam is essentially free. So, in the long run, you save on operating cost because you have no fuel expense; the electricity becomes much cheaper compared to that generated by coal or oil.

One disadvantage of geothermal is that you have to develop it where you find it. In other words, it is site-specific; you cannot pipe and store the steam then sell it somewhere else. You have to convert it first into electricity before you can transport it through submarine cables and transmission lines. You also cannot build single units of large-capacity power plants because that will require more steam and therefore longer pipelines to get the steam. But because it constantly changes in temperature, steam cannot be transported through a five-kilometer pipeline, for instance, without it turning into water before reaching the end of the line. That is why it is not feasible to build single-unit, large-capacity power plants over geothermal wells; on the average, you would need eight geothermal wells to build a 100-megawatt power plant. What PNOC initially did in Tongonan, for example, was to build three 37.5-megawatt geothermal plants that could alternately supply the transmission grid with electricity.

Another thing we must remember is that, if not done properly, geothermal development can have adverse environmental impact. The biggest pollutant is the water that comes from underground. What comes out of the geothermal well is a combination of steam and water. We separated these, with the steam passed on to the geothermal turbine to generate electricity, and the water re-injected to the well. We actually pioneered the use of this re-injection method, at least in the country. Since the water contained high levels of boron, it was very dangerous to discharge it into the river system as this would flow to the rice fields.

PNOC-EDC was actually the first Filipino company to set up in the late Seventies its own environmental protection division, whose function was to ensure that our operations people were not polluting the environment. They were prohibited from discharging into the river the water extracted from the geothermal fluid, since it would raise toxic levels beyond those allowed by the World Health Organization. PNOC-EDC's environmental division constantly sampled the water adjacent to the geothermal field and if found to contain significant amounts of boron or other toxic material, the geothermal field was shut down until appropriate remedies were taken.

Several years later, I learned of another potential environmental hazard that we needed to keep in mind. In August 1985, I was given the honor to keynote the international geothermal convention in Hawaii. On that occasion, a group of scientists from the Smithsonian Institute talked to me about our geothermal development program. Although they praised our accomplishments, they also warned that we should seriously consider the possible after-effects of accelerated utilization of more than 1,200 megawatts in a matter of twelve years (we had already tapped nearly 900 megawatts by then). A major concern of the scientists was the possible increase in the frequency of earthquakes, although no exact scientific data could support such a theory. Just the same, the thought of earthquakes occurring because of geothermal exploration scared me, and for a split-second I imagined people blaming me if the scientists' fears came to pass.

Upon my return to Manila, I talked to our geothermal team about the American scientists' concern. I learned that the issue was not exactly unknown to our staff, since the New Zealand Geothermal Institute that PNOC-EDC worked with was also studying the same. My own personal belief was that energy exploration and development activities (especially geothermal) tamper with nature and this cannot be helped; we simply had to be aware of the potential hazards and do our best to minimize whatever adverse impact these activities may have on the environment. And despite the lack of definitive scientific data to support the concern about potential earthquakes, I cautioned the PNOC-EDC staff that we should perhaps level off tapping additional geothermal potentials until after a few years of operation.

When we began the country's geothermal development program in 1974, its primary objectives were to support the expansion and growth of the power program, and to reduce the share of crude oil in the power mix. Substantial financial resources were mobilized for the program. From 1976 to 1983, PNOC-EDC's Geothermal Division spent over Php 1.8 billion, of which Php 720.7 million came from the Oil Industry Special Fund. The investment definitely paid off because, by 1980, the various geothermal fields generated a combined power output of 446 megawatts, making the Philippines the second largest geothermal energy producer in the world behind the United States (see Table 2.2: Performance Indicators in the Geothermal Sector). Additional power units constructed in Tiwi (Albay), Bacon-Manito (Sorsogon), Tongonan (Leyte), and Negros Oriental in the succeeding years brought the total geothermal power generating capacity to 892.5 megawatts at the end of 1984. Some private companies like Caltex (Philippines) and Total Exploration decided to participate in the geothermal

Table 2.2
Performance Indicators in the Geothermal Sector

Year	Cumulative number of fields	Number of wells	Footage	Proven additional field capacity (MW)	Cumulative power production (MW)
1973	1	4	23,834	10.4	—
1974	2	4	22,203	17.3	—
1975	2	8	48,556	28.5	—
1976	3	18	91,116	120.7	—
1977	3	24	143,948	131.5	3
1978	3	34	183,156	191.6	58
1979	4	58	317,496	315.0	223
1980	4	59	336,625	385.0	446
Total	4	209	1,166,934	1,200.0	446
Average (1973-1976)	2	9	46,427	44.0	—
Average (1977-1980)	4	44	245,306	256.0	110

Source: The National Energy Program, 1981-86. Makati: Ministry of Energy. Republic of the Philippines

program in 1982. More important, our geothermal output substantially contributed in reducing dependence on imported crude. For instance, geothermal energy displaced US$ 94 million or Php 940 million worth of oil imports in the first half of 1983. At the height of the debt crisis in 1984, geothermal energy's 894-megawatt output accounted for 8.33 percent of total energy consumption (the second-largest non-oil source after hydroelectricity, which provided 10 percent) and displaced 7.8 million barrels of fuel oil equivalent worth about US$ 227 million.[41]

For his contributions to geothermal development, Dr. Arturo P. Alcaraz won the Ramon Magsaysay Award for Government Service in 1982. Finally, his lifelong work had been fully recognized.

Coal Development

Coal as an energy source had been used by small shops and foundries in the Philippines as far back as the 1600s. Then, with the introduction of first locomotives by the Manila Railroad Company (now Philippine National Railways), coal was also used as transport fuel. In the Thirties, Cebu Portland Cement Company and Rizal Cement Company also used coal to feed their cement plants. On the whole, coal was a principal energy source in the country before the Second World War and through the early postwar years, until cheap oil gradually displaced this resource such that, by the middle of the Sixties, coal accounted for an insignificant 0.1 percent of national energy demand.[42] But with the oil crisis of 1973-74, the development and utilization of coal became important in the government's energy diversification objective.

Being the only known fossil fuel to exist in commercially significant quantities in the Philippines, coal development and utilization was a logical strategy in energy diversification. Toward this end, the government enacted PD 972, known as the "Coal Development Act of 1976,"[43] and designated

---

41. See my article, "Energy Self-Reliance and the Debt Crisis," which first appeared in the *Fookien Times Philippines Yearbook 1984-1985*, and subsequently published by Energy Forum (a publication of the Philippine National Oil Company), Number One, 1985, 6.
42. "The Great Coal Revival," Energy Forum, 17.
43. Presidential Decree No. 972, "Coal Development Act of 1976," July 28, 1976.

the Bureau of Energy Development as lead agency. The new coal law aimed to strengthen the coal mining industry and to improve mining methods by mandating traditional coal miners to pool their resources and work in partnership with government in exploring and developing coal mines. Small-scale coal miners were also required to unify their coal landholdings into sizeable coal blocks of at least 1,000 hectares each, and enter into new operating contracts with the government. The new coal operating contract system was patterned after the service contract for oil exploration.

Coal exploration was another virgin territory for PNOC. Until 1976, coal mining was undertaken by small-scale miners who did not have the resources to upgrade their capabilities. They used a very primitive process of pick-and-shovel methods, which tended to be wasteful, was grossly unsafe, and yielded unsatisfactory results. The activity was also confined primarily to Cebu and Zamboanga. In those days, the sheer absence of sustained, serious work by the government and private sector in such important areas appalled me. Each night, as I pored over the statistics on coal, I gritted my teeth in both frustration and anticipation. Surveys indicated a large potential reserve of 1.4 billion metric tons. My calculator told me that this was equivalent to four billion barrels of oil. Of course, not all of this potential reserve could be mined, since you had to assume that about half would be contaminated. Nevertheless, the prospect of mining the equivalent of two billion barrels of oil was enough incentive for us to pursue the development of coal as an alternative energy source. Really, even the potential equivalent of a billion barrels of oil was more than sufficient to motivate us. And I thought that there were large potential customers. For example, the country's cement plants needed a cheaper alternative to crude oil. But to convince them to shift to coal, they would have to have an assured supply of quality coal, delivered at their outlying locations at reasonable prices. In other words, we had to build the coal industry from scratch, and this meant developing both supply and demand, as well as building the infrastructure that would serve both ends of the industry. We targeted that our immediate market would be the power plants and the cement industry, which were primary users of fuel oil.

To get us started, I assigned Mario V. Tiaoqui to head PNOC-EDC's Coal Division. Mario was an assistant manager at the Bataan Refining Corporation when PNOC took over Esso (Philippines). I remember the American executives telling me that Mario was among the Filipino staff who showed great potential to become an executive. I decided to make him the first head of Petron and, subsequently, we thought that he should get involved in energy development. Eventually, he headed the coal division, which was tasked to implement PD 972 by introducing scientific and mechanized mining methods in a coal block in Uling, Cebu.

When we showed the viability of large-scale operations, the small-scale, primitive, and backyard miners became convinced in combining their resources and efforts, adopting more efficient methods, and collectively developing the coal block. We provided technical assistance whenever necessary. Coal exploration took a dynamic turn in 1977, when production reached 284,000 tons and proven reserves reached 28 million tons. Coal production for that year, though still modest for the country's growing energy demand, displaced about 754,000 barrels of oil and generated savings on imported crude worth US$ 12 million.

That initial effort set us on the road to a comprehensive coal development program. In 1979, I spun off two new companies. With a loan of US$ 14 million from the Asian Development Bank, we formed the Malangas Coal Corporation that would take charge of extracting four million tons of high-grade bituminous coal from the Diplahan Colliery in Malangas, Zamboanga del Sur. The Malangas coalmine was the first really intensive and technologically advanced mining of coal in the Philippines.

The other firm, PNOC Coal Corporation, exploited reserves in the Little Baguio Colliery (also in Malangas), and new coal areas in Bislig (Surigao), Calatrava (Negros Occidental), Clarin (Bohol), and Uling (Cebu). By 1980, we were able to raise coal production's output to 328,786 metric tons, nearly eight times more than the 1973 production of only 39,004 metric tons; of the total production, Cebu accounted for 65 percent and the rest contributed by Malangas (Zamboanga del Sur), Polillo (Quezon), Batan

Island (off Albay), and Semirara Island (off Antique)[44] (see Table 2.3: Performance Indicators in the Coal Sector). PNOC also conducted feasibility studies on the conversion of eighteen factories and several power plants to coal use. By 1984, PNOC Coal Corporation (PNOC-CC) and Malangas Coal Corporation were operating seven coal-producing mines with a combined output of 274,000 metric tons; together with the production of its contract coalmines, PNOC accounted for 28 percent of the national coal output.[45]

In due time, the accelerated coal development program became an important component of the government's thrust toward energy self-sufficiency. An integral aspect of this program was the conversion of the fuel base of selected industries, especially cement, from petroleum to coal. Now, unlike geothermal that is site-specific, coal can be transported from the site to the end-user. To ensure the delivery of coal to our intended clients, we had to establish a nationwide coal logistics system consisting of an infrastructure network of coal terminals, ports, relay stations, and blending and off-loading facilities. We also planned to upgrade the rail lines of the Philippine National Railways so that we would have a modern railway system from Batangas to Manila that transported not only coal but also passengers.

To oversee the construction of these infrastructure facilities and regulate downstream activities, the National Coal Authority (NCA) was created by PD 1722[46] in September 1980. Besides ensuring the quality of marketable coal, the NCA was tasked to develop a program for logistical facilities and infrastructure for the industry, and to coordinate as well as monitor the actual supply and demand situation in the coal market that might require exploration to fill in shortfalls. To finance the installation of a logistics and supply network, Letter of Instruction (LOI) 1159 appropriated Php 150 million from the Oil Industry Special Fund. In 1983, coal receiving and blending terminals were completed in Poro (La Union)

---

44. *The National Energy Program*, 18.
45. Philippine National Oil Company, Annual Report 1984.
46. Presidential Decree 1722, "The Charter of the National Coal Authority," September 16, 1980.

Table 2.3
Performance Indicators in the Coal Sector

Year	Cumulative number of service contracts	Production (MT)	Reserves (In million MT)	
			Proven	Potential
1973	—	39,004	77	125
1974	—	50,745	77	125
1975	—	105,128	77	125
1976	—	122,000	77	125
1977	5	284,554	77	125
1978	17	254,515	77	125
1979	37	263,393	125	1,010
1980	45	328,786	186	1,690
Total	45	1,448,125	186	1,690
Average (1973-1976)	5*	79,219	—	—
Average (1977-1980)	10	283,000	27	390

*Started 1976

Source: The National Energy Program, 1981-86. Makati: Ministry of Energy. Republic of the Philippines

and Iligan (Lanao del Norte), while an interim terminal was put up in Batangas. The NCA assigned PNOC Coal Corporation to provide for inter-island coal movement, for which PNOC commissioned two specially designed marine vessels. I appointed Mario Tiaoqui as concurrent executive director of NCA.

In trying to convince the cement companies to pilot coal-conversion programs, the Ministry of Industry and the Development Bank of the Philippines (DBP) offered an assistance package consisting of a common supplier for the conversion equipment and a financial scheme with reasonable terms. Industry Minister Roberto Ongpin conceptualized this assistance package, and he purposely coursed the financial assistance through DBP to avoid corruption on the part of the supplier. In addition, the Ministry of Industry, together with the Ministry of Energy, guaranteed the plants that the price of coal from PNOC-operated and contract mines would not exceed 65 percent of industrial fuel oil price; any price differential above 65 percent would be shouldered by the Ministry of Energy. The cement plants' conversion process was slow, but we eventually realized our goal. In 1982, only six plants shifted from oil to coal; by 1984, all seventeen cement plants had converted.[47] Considering that we practically started from zero, PNOC's achievements in coal development were nothing short of impressive, and I give credit to Mario Tiaoqui for having spearheaded this endeavor as concurrent head of the National Coal Authority and the PNOC Coal Corporation. Mario more than met the challenge. I also want to add that under the current structure of government today, it would be impossible for one person to head both a regulatory agency (NCA) and a business corporation (PNOC-CC). But in those days, it was convenient to do this and I will submit that it remains a convenient arrangement for present purposes.

Unfortunately, the economic crisis of 1983 forced us to suspend the construction of a terminal receiving facility in Batangas, and to put off the plan of building a railway system from Batangas to Manila. Then, in 1984, as the country implemented the International Monetary Fund's structural

---

47. Philippine National Oil Company, Annual Report 1982 and Annual Report 1984.

adjustment program, we had to allow market forces to operate in the coal industry. This meant, among others, that the cement plants were free to buy coal from other suppliers, and the joint NCA-PNOC coal price guarantee had to be withdrawn. Despite some limitations, the coal development program did result in substantially increased coal production; from 39,000 metric tons in 1973, the national coal output reached 1.2 million metric tons in 1984. It also resulted in the conversion of seventeen cement plants and two NPC oil-fired power plants.

The coal development program also yielded valuable information on the quality of our coal. Most of our coal deposits are lignites—very young coal with low heating value, as measured by the British Thermal Unit (BTU). Because of that, we needed to mine greater volumes of coal if we were to use it for generating electricity; another option was to construct the power plant near the mine mouth. But while we do have very good coal deposits, such as those in Zamboanga and Cebu, they are rather small and do not grow big coal seams, which makes them more expensive to mine. Despite these limitations, I am proud to say that many of the cement plants that joined our conversion program still use coal. More important, coal continues to account for about 25-30 percent of the power sector's energy mix.

I recognize that, at present, the use of coal as energy source is under severe question due to environmental concerns. Even during our time, environmentalist groups were already calling attention to the adverse impact of using coal, and we were not oblivious to these issues. However, we were confronted by the need to diversify our energy sources, and the wide availability of coal made it feasible for PNOC to develop this resource on a commercial scale. At the same time, from an economic point of view, installing a coal-fed power plant was cheaper than a diesel or bunker oil-fed power plant. This means that electricity from a coal-fired power plant is also much cheaper per kilowatt. There is really a trade-off in the relationship between economics and environment.

In general, energy and environment contradict one another, especially with regard to coal. There is no doubt that coal contains injurious elements that, when released through the exhaust of coal plants, can affect the immediate environment. In fact, a number of energy substitutes that we were developing were potentially harmful to the environment but especially coal, since ours is mostly lignite coal—the lowest form and also the most corrosive. But I was assured by manufacturers of coal equipment for power generation that they were putting the necessary safeguards that would minimize sulfur emission once coal was fed into a boiler. These safeguards were adopted as matter of policy.

Subsequent advances in technology have made environmental hazards relatively more manageable. Coal plants that were built in the Nineties actually meet current clean air standards. The oil prices today will show that you really have to develop substitutes for oil lest it be very costly for power generation.

Nonconventional Energy Sources

The inclusion of nonconventional energy (or "non-con," as we called it then) among our future sources was a novel feature of the country's energy source mix. Though initially small, the projection showed its contribution to be rising rapidly to approximate the equivalent of three million barrels of oil. Identified as primary contributors to "non-con" were alcogas, coco-diesel, gasifiers, biogas, and solar wind. Along this line, the Ministry of Energy established the Center for Non-Conventional Energy Development to undertake research and the propagation of indigenous alternative energy devices using solar radiation, wind, biogas, and agricultural wastes. I want to focus on two major programs that we carried out, *alcogas* and *coco-diesel*, because our experience in these programs provide important lessons that may be useful for our present purposes, since these programs are becoming fashionable again in the face of constantly rising petroleum prices.

Alcogas. The demand for energy self-sufficiency necessitated that we look into ways by which we could create new uses out of certain agricultural crops, like sugar, to produce new types of fuel that would reduce our gasoline consumption. That was what we hoped to achieve when we went into alcogas. The program was designed not only to provide alternative transport fuel but to generate employment opportunities in the rural areas as well. In particular, since the alcogas program was based initially on molasses, one of the by-products of the sugar-refining process, we thought that we would be helping the sugar industry as well.

A few years before we formally embarked on an alcogas program, I had gone on an official trip to Brazil to study the alcohol potential, since this country pioneered the use of alcohol for their fuel. I remember that during our visit in 1976, some quarters were laughing at Brazil for trying to make alcogas. There were jokes like, "Alcohol is for drinking, not for cars!" Yet, Brazil pursued its program and never tired of innovating and improving on its alcohol production. Brazil also developed its own ethanol-based motor engines and cars. Today, Brazil is the world's largest producer of alcohol and everyone praises the country for its "super-efficient" production system.

The Philippines had actually been an alcohol producer since sugar centrals were established by American pioneers. The process involved bringing the sugar canes to the sugar central for refining; as a by-product, molasses was produced, from which alcohol was extracted. But we did not have much use for alcohol then except in the production of rhum, so using it with gasoline seemed an efficient and productive way of utilizing alcohol. However, to obtain cheaper alcohol with less water content, we would have to produce alcohol directly from sugar cane without having to pass the sugar-refining process. The financial requirements for such, however, did not seem justified.

Then, one day, my attention was called by a report conducted by the World Health Organization (WHO), stating that the permissible lead and mercury content found in the average Filipino child living in congested areas was six times more than the international permissible standards. The WHO report pointed to the use of tetraethyl lead (TEL), which we used as an octane enhancer in our refineries' gasoline production, as the primary cause of the high mercury and lead content among children. Subsequently, we obtained a grant from the US Agency for International Development to seriously study possible substitutes for TEL, and alcogas turned out to have the greatest potential.

We tapped the Institute of Biotechnology at UP Los Baños (UPLB), and Dr. Emil Q. Javier, who was then chancellor, was most enthusiastic in his estimation to substitute alcohol for TEL. At the time, the institute had many scientists but they were largely confined to teaching. I think, because they also had obligations to the university (similar to Art Sali's situation when he returned from postgraduate studies in England), they had no recourse but to stay in UP. It so happened that Emil's passion then was to try and hasten the propagation of mongo beans through biotechnology. I told Emil, "Why don't we do this research on alcogas? That way, we can help your scientists." He agreed and so I placed around twenty of the institute's scientists, led by Dr. Elpidio "Pids" L. Rosario, on PNOC's payroll to work on the alcogas research, looking at how to shorten the process so that we could plant sugar cane and produce alcohol without having to convert the cane juice into refined sugar.

The rising costs of gasoline also caused PNOC and the Progressive Car Manufacturing Program to renew interest in alcohol fuels in 1976. A series of tests were conducted with locally produced cars, which established the unsuitability of hydrous alcohol (95 percent pure) as a blending material with gasoline. I remember Pids Rosario briefing me about the study, which showed that to displace portions of gasoline, the alcohol should be 99.5 percent pure or anhydrous. With less than 0.5 percent water, alcohol and gasoline are miscible in all proportions. In the United States, "gasohol" (a blend of gasoline and ten percent anhydrous alcohol) was commercially distributed in 1977. In Brazil, the initial alcohol program produced a 20 percent anhydrous alcohol blend. Based on the tests and other countries' experience, we planned for a National Fuel Alcohol Program that was targeted to produce anhydrous alcohol of at least 99.5 percent purity.

The alcohol study eventually led to the formal establishment of the National Institute of Biotechnology and Applied Microbiology (Biotech) at UPLB. It was set up in 1979 to undertake research in the fields of genetics, microbiology, chemistry, and engineering. In time, Biotech moved from studies on fuel-alcohol production microorganisms to areas relating to continuous flow fermentation of sugarcane molasses, and conversion of cassava into alcohol. I have a story to relate in this regard.

When I presented to President Marcos the alcogas research project in Emil Javier's presence, I mentioned in passing that this biotechnology effort, although geared specifically toward helping government to produce alcogas efficiently, was in tune with the First World's attempt at DNA research. The president, being a science buff and very much fascinated by biotechnology himself, got excited and ordered me to contact two California-based companies pioneering in DNA technology—the Cetus Corporation of Berkeley and the Genentech Corporation of San Mateo—in the hope of upscaling the research efforts of both PNOC and UPLB. When I half-jokingly asked the president if he could fund the Institute of Biotechnology, he immediately issued instructions to his purse-keepers and, in a few months' time, Emil Javier received a donation of Php 20 million from the Marcos Foundation on top of Php 8.8 million from the Oil Industry Special Fund to cover initial financial requirements.

As we were trying to learn more about alcogas, the government set up an interagency task force that would look into the organizational, capital, and other resource requirements for a national alcogas program. In July 1979, the task force came up with the following report on a proposed National Alcohol Fuel Plan:

- The Program
  - Immediate Targets
    1. Anhydrous alcohol blended with regular and premium gasoline with a maximum of 15 percent alcohol by 1985
    2. Blending at bulk plants of oil companies
    3. Expected date to attain 15 percent alcohol blend – 1985
  - Long-Term Target – hydrous alcohol for 100 percent alcohol engines

- Alcohol Distilleries
  The alcohol requirements of the Alcogas Program were expected to be produced from three basic models of distilleries:

➤ Model I – Small Annexed Distilleries
   Existing or new distilleries annexed to existing sugar centrals with capacities ranging from 30,000 to 60,000 liters per day. This was for immediate implementation and intended to open opportunities for existing sugar centrals. The feedstock could be sourced from incremental cane production or from molasses produced in the cane district.

Projected Capital Investment Costs: Php 732 million for ten units
Estimated Agricultural Financial Requirements: Php 332 million

➤ Model II—Large Annexed or Autonomous Distilleries
   These distilleries would have capacities ranging from 120,000 to 240,000 liters per day, were intended to supply major demand areas like Metro Manila, and were expected to provide favorable economies of scale in both investments and operating costs for agricultural and distillery operations. The autonomous distilleries were envisioned to have their own supply of canes and would process canes directly to alcohol.

Projected Capital Investment Costs: Php 6.668 billion for twelve units
Estimated Agricultural Financial Requirements: Php 2.3 billion

➤ Model III—Autonomous Municipal Distilleries
   These distilleries were to have capacities ranging from 30,000 to 60,000 liters per day and were envisioned to supple regional local area requirements. Some of the perceived advantages of this model were: lesser vulnerability of the supply system due to diversification of production units and minimization of the effects of regional climatic variation; reduction of transport costs of raw material; and, simpler utilization of distillery waste.

Projected Capital Investment Costs: Php 3.341 billion for twenty-five units
Estimated Agricultural Financial Requirements: Php 2.3 billion

The projected alcohol production if all the distilleries were in full operation was 925 million liters per year by 1988. The area required for feedstocks were 211,500 hectares for sugar cane and 51,400 hectares for cassava.

The above report by the interagency task force eventually became the basis for President Marcos to issue Executive Order 580,[48] creating the Philippine National Alcohol Commission (PNAC) in 1980. In the same year, we organized the PNOC Alcohol Corporation (PNOC-AC) and subsequently teamed up with PNAC in producing and developing fuel alcohol or alcogas, which would be a blend consisting of 85 percent gasoline and 15 percent alcohol. But, to produce anhydrous or water-free alcohol suitable for blending with gasoline into alcogas, we needed dehydration facilities. The government, through LOI 983, released Php 10 million from the Oil Industry Special Fund to PNOC, Php 7.5 million of which was loaned to Victorias Milling Company in Manapla, Negros Occidental—the heart of the sugar industry—to finance the installation of the dehydration plant. We used the balance to purchase other logistical equipment as well as for PNOC-AC's working capital requirement. The assistance enabled the company to launch alcogas for commercial use in Negros Occidental in 1980.

In the succeeding years, the alcogas program expanded to Negros Oriental and Panay Island, necessitating an increased production of anhydrous alcohol. Further financial assistance was needed for the installation of additional dehydration facilities. Accordingly, Php 4.5 million was released from the Special Fund to PNOC under LOI 1153, and lent to Central Azucarera de la Carlota for the upgrading of its distillery.

Looking back, we started on fire with this one, only to come up short of our expectations because we did not (and could not) see that the sugar industry was going to collapse within the next five years. We were carried away by the momentum of our string of successes in the early years. When we formed PNOC-AC in 1980, we felt nothing was going to stop us. The ambitious program called for the displacement of at least 15 percent of gasoline requirements within the decade. To achieve this, we planned to install ten distilleries with a capacity of 180,000 liters per day. That is why we entered into a long-term supply contract for anhydrous or water-free alcohol with Victorias Milling, whose distillery was rehabilitated

---

48. Executive Order 580, "Creating an Alcohol Commission," February 16, 1980.

through our financial assistance. We dared to rehabilitate the sugar industry, which was already on its downward cycle.

The alcogas program was one of those learning experiences for PNOC and the Ministry of Energy. It cost a lot to install a distillery, and it cost even more when it had to be equipped with dehydrants. This made the production of one liter of water-free alcohol even more expensive than to produce one liter of gasoline. In addition, we found out that because of nearly a century (or more) of overexploitation, most of Negros's sugar cane fields had declined in quality and could not produce the appropriate sugarcane that we needed for the program. This meant that, to achieve our goal of 15 percent displacement of premium gasoline by alcohol within a ten-year period, we needed about 250,000 hectares of sugarcane fields. But, as I reported to the president, we could not "manufacture" such a plantation overnight; even Dole's pineapple plantation of 10,000 hectares took seven years to develop from the time of project conception.[49] The costs simply outweighed the potential benefits. This, plus the easing of oil prices, made us go slow on the program.

But I will not be too harsh in assessing the alcogas program. Of course, at the time, we were so frustrated. But in retrospect, we were doing the spade work and finding out in the process that we would need proficiency, knowledge, foresight and, most important of all, appropriate sugar plantations. The alcogas program belonged to an entirely different field—it was largely agricultural rather than energy-related. Also, alcohol cannot be produced overnight; the whole process from planting sugar canes to utilizing the alcohol would take about ten years, thereby requiring economies of scale in order to recoup the huge capital investment. Unfortunately, the Aquino government dismantled the program, thereby depriving the country of an opportunity to deepen its knowledge and experience in developing alcohol for fuel.

The other benefit of the alcogas program was that it provided us with another opportunity to tap our local scientists and encourage them to direct

---

49. Geronimo Z. Velasco, "Organizational Requirements for an Accelerated Alcogas Program," July 5, 1979 (Memorandum to President Ferdinand E. Marcos).

their research in areas with potential industrial application. I actually liked working with the UPLB Biotech's scientists. They were very good, very talented, but they did not have business sense which would have brought them places. They were also very shy and inarticulate, and Emil Javier was actually the only scientist I met at the time who was both very good and very articulate. I think that was the reason why he emerged as the de facto leader of UPLB's community of scientists. Eventually, I recommended him to become Minister of Science. Ten years later, President Fidel Ramos appointed him as president of UP. Unfortunately for the institute we built at Los Baños, it became a casualty of politics when a new government came in, given the institute's prominent association with the ministry and the Marcos administration. It regained momentum when Emil Javier became UP president in 1993.

You are probably wondering how I got into this conscious effort at working with scientists. I guess part of it stems from a genuine admiration for what they do, and partly because my engineering background helped to create in me a basic appreciation for science. And in working with them, I think we unwittingly created a symbiotic relationship. The scientists possessed the basic knowledge and theories, but most of them did not have much exposure in terms of application. That is where I came in—through PNOC and the ministry's various energy development programs, I provided a venue for them to test the applicability of their ideas. Actually, at the time, two ministries were working closely with UP's scientists; the other was the Ministry of Agriculture under Arturo Tanco, through the Institute of Plant Breeding at UPLB.

Coco-diesel. In 1981, the coconut industry was seriously affected by the depressed prices of crude coconut oil (copra) due to an oversupply in the world market and a domestic surplus that could not be accommodated by the local market. To protect the industry, the president directed the electric power generation and land transportation sectors to partially shift to coco-diesel or coco-fuel oil blend. To keep the coconut industry viable, a total of 48.8 million liters of coconut oil were purchased for use as industrial fuel extender, at a cost higher than petroleum-based fuel. The cost differential was covered by an appropriation of Php 145.1 million under LOI 1167. An additional Php 12 million was released from the Oil Industry Special Fund (also through LOI 1167) as price assistance to program participants to preclude financial loss on their part.

Unfortunately, much like alcogas, the coco-diesel did not perform effectively as a fuel substitute. The quality of our coconut oil was such that it tended to coagulate fast, so it could not blend fully with diesel. Many transport operators who tried out coco-diesel were incensed because they had to use a toothbrush, literally, to rid the engine of coagulated coconut oil. They were saying, *"'Yang project ni Velasco, kelangang sipilyuhin!* (You need a toothbrush to make Velasco's project work!)."

Coco-diesel can be used but you really have to process it well because there are many impurities in coconut oil, which is why it coagulates fast. It is also waxy, which is not good for the engine. So the use of coconut oil as a fuel blend can still be done, but it has to be well-planned and well-studied. Like alcogas, it cannot be produced overnight. And you will only know what else needs to be done when you have already started the processing. I think we can safely say that we are still about five years away from producing the right coco-diesel to run our transportation engines.

Nonconventional Energy Research. Foreign ambassadors managed to convince me that if we could show sincere and serious efforts in pursuing the applicability and adaptability of our nonconventional energy resources like biogas, biomass, and solar applications, their respective governments would be willing to fund part of our research efforts. It was on this basis that we established the PNOC Energy Research Center.

With PNOC being essentially a business organization with strong management expertise and with a potential capability to tie up with the academe, I discussed the possibility of establishing a research center with then-UP President Edgardo J. Angara, against a donation of Php 15 million that UP could use in constructing a new building for its Institute of Geology. Angara's dilemma was that the German government was donating laboratory facilities and equipment to the institute, but on condition that UP provide the building to house these. So when he called me up and I broached to him the quid pro quo, he immediately agreed to donate one hectare of UP's property for PNOC's research center.

Even from a business standpoint, I thought the deal was excellent! We would be hitting two birds with one stone. The availability of geology graduates in the Philippines was fast declining. The construction of the building allowed the Institute of Geology not only to have better facilities, but also to accept more students. On the other hand, by constructing a research energy building in the UP Diliman area and in the process tapping the expertise of the College of Engineering, we were able to convince foreign governments of our seriousness in pursuing the search for nonconventional energy. This line of reasoning convinced an initially reluctant President Marcos to release Php 15 million from the Oil Industry Special Fund for the construction of the Institute of Geology's building.[50]

Shipping and Transport

In addition to the energy development projects, PNOC engaged in building tankers and operating a shipyard, since it was eventually determined that the oil industry needed substantial assistance in upgrading and updating its oil transport capability not only for current needs but also in anticipation of future growth in the country's oil consumption. In justifying PNOC's venture into shipyard operation, we believed that unless a new transport fleet was pressed into service, the industry would have difficulty delivering oil and other petroleum products throughout the country. At the same time, the various bulk plants located in outlying areas would need additional storage facilities to accommodate larger inventories in the event of an increased demand for oil. There was no doubt that the facilities PNOC built could meet these eventualities.

The PNOC Shipping and Transport Corporation had actually been operating an aging, if not aged, fleet composed of small tankers (with a capacity of about 5,000 tons) and self-propelled oil barges. These tankers and barges were acquired when the Luzon Stevedoring Corporation (LUSTEVECO) was integrated into the PNOC group. Given the quality of

---

50. The Institute was later renamed the National Institute of Geological Sciences.

the existing fleet, the only way we could upgrade the system was for PNOC itself to build and develop a shipyard operation.

Capt. Pablo "Pete" N. Sare, originally LUSTEVECO's vice president for engineering, designed the new facilities that included a number of "firsts" in the country. Foremost among these was a hydraulically operated dock-lift that was capable of lifting vessels of up to 20,000-ton deadweight. The dock-lift was integrated into a side transfer system, which had an initial capacity for six ships in dry berths. The said system was the first and the biggest of its kind built in the Philippines, and I give credit to Pete Sare, who served for a number of years as a member of the London-based Technical Committee for Asia of the prestigious Lloyd's Register of Shipping.

Upon establishment of the shipyard facilities, we built four 5,000-ton deadweight tankers. These vessels were designed, built, and classed according to American Bureau of Shipping rules for transport of petroleum products, and were the first ships built in the country that were consistent with the regulations of the International Classification Society.

The knowledge and experience gained by our shipyard personnel in the modern method of building ships in blocks, coupled with the acquisition of technology in the design and preparation of working drawings, could have been the foundation of a shipbuilding industry in the country. However, PNOC's venture became an object of criticism of the Asian Development Bank (ADB), whose consultant, Arthur D. Little Associates, observed in its evaluation report that PNOC should have encouraged the other small shipyard facilities operating in the Philippines to build a new and modern fleet. But it was precisely because the local shipyard industry did not have the capability to build the type and quality of tankers needed by the oil industry that we rationalized the establishment of PNOC's shipyard facilities. We had even planned to upgrade the various bulk plants throughout the country, and PNOC could use the shipyard for building these bulk storage facilities.

PNOC had never intended to monopolize shipyard operations. In fact, it had sold another property adjacent to the location of the PNOC shipyard to Keppel Shipyard, a major shipyard manufacturing facility owned and controlled by the Singapore government. Keppel Shipyard subsequently established a yard in the same property. In my discussions with ADB's representatives, I argued that, how come they were not objecting to the Singapore government being involved in the operation of a shipyard facility, and yet PNOC—which had a direct interest in the upgrading of its fleet—was being criticized for effectively infringing on "private interests"?

The "total energy program" that PNOC and the Ministry of Energy implemented from 1976 to 1983 paid off in terms of reducing the Philippines' dependence on oil imports by around 44 percent by 1984, a track record that was often cited internationally. The National Energy Program (1981-1986) launched by the country shortly after the outbreak of the second oil crisis in 1979-1980 was praised by the World Bank as a model for Third World countries similarly situated. The combination of increased use of indigenous resources and lower average oil import cost led to a reduction in our oil import bill, which helped the economy at a time when it was already experiencing severe financial crisis (see Figure 2.1 and Figure 2.2). Although some of these alternative-energy programs did not produce the anticipated fuel substitutes, these programs nonetheless enhanced our knowledge in terms of both the potentials and limitations of indigenous resources, providing valuable lessons that remain important to this day as guide to government's energy planning.

For PNOC, the shift to total energy was instrumental in achieving what is probably one of the most rapid growth rates a company has ever had. In a span of six years, from 1974 to 1980, we created twenty-three subsidiary companies that focused on three areas: petroleum refining and marketing; transport and logistics; and, energy exploration and development (see List of PNOC Subsidiaries on pages 83 to 85). To ensure that these subsidiary companies were properly coordinated, I created a management executive committee that would oversee everyday operations. Antonio V. del Rosario, PNOC senior vice president, headed this committee as executive director.

Again, the creation of these companies in such a short time would not have been possible were it not for martial law. Decision making within the PNOC's board of directors was primarily organized around myself, Defense Secretary Juan Ponce Enrile, and Executive Secretary Alejandro

Chapter 2: Working for Energy Self-Reliance • 79

Figure 2.1
Primary Energy Consumption, 1974-1983
(Million Barrels of Fuel Oil Equivalent)

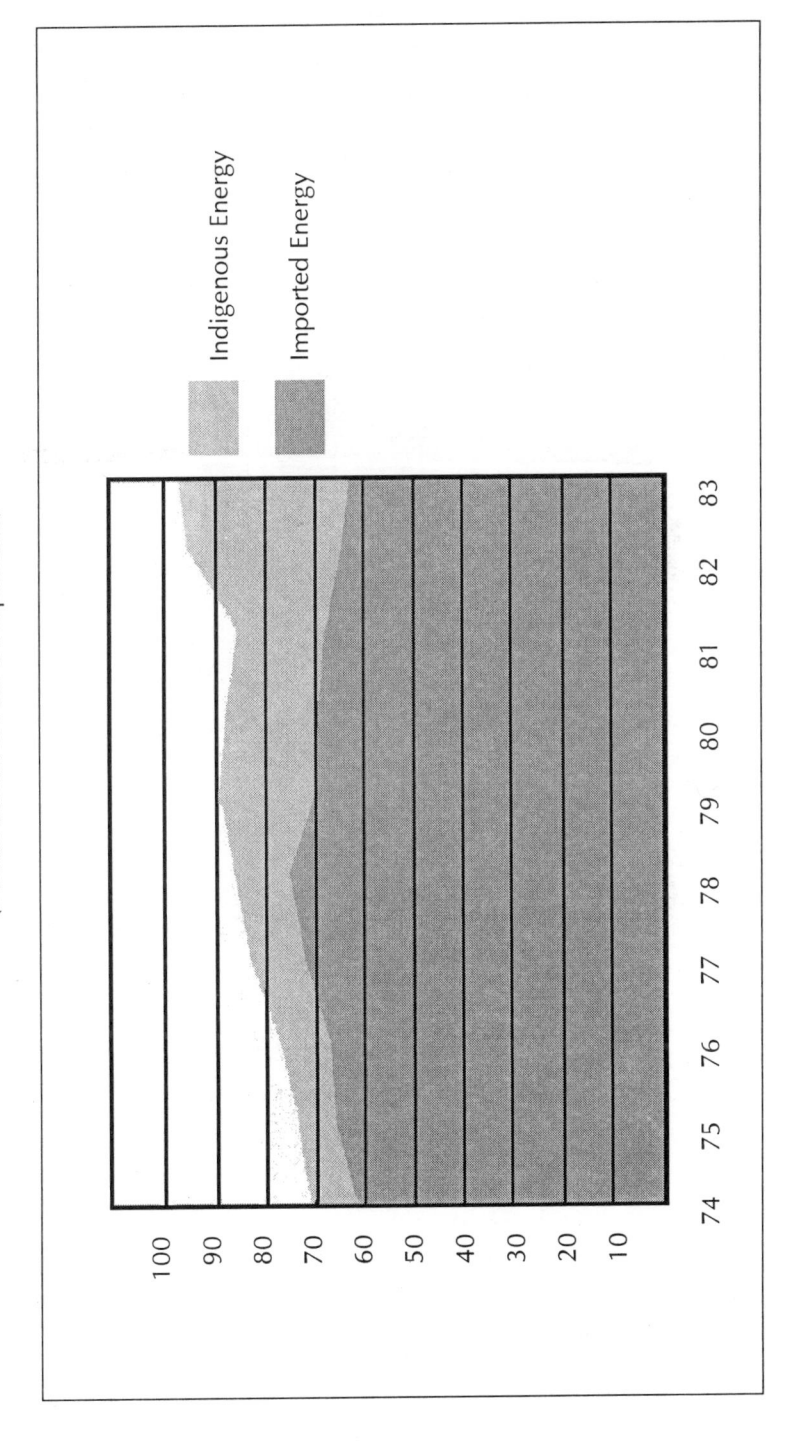

■ Indigenous Energy
■ Imported Energy

**Figure 2.2
Average Crude Oil Import Cost, 1974-1983
(Per Barrel [FOB])**

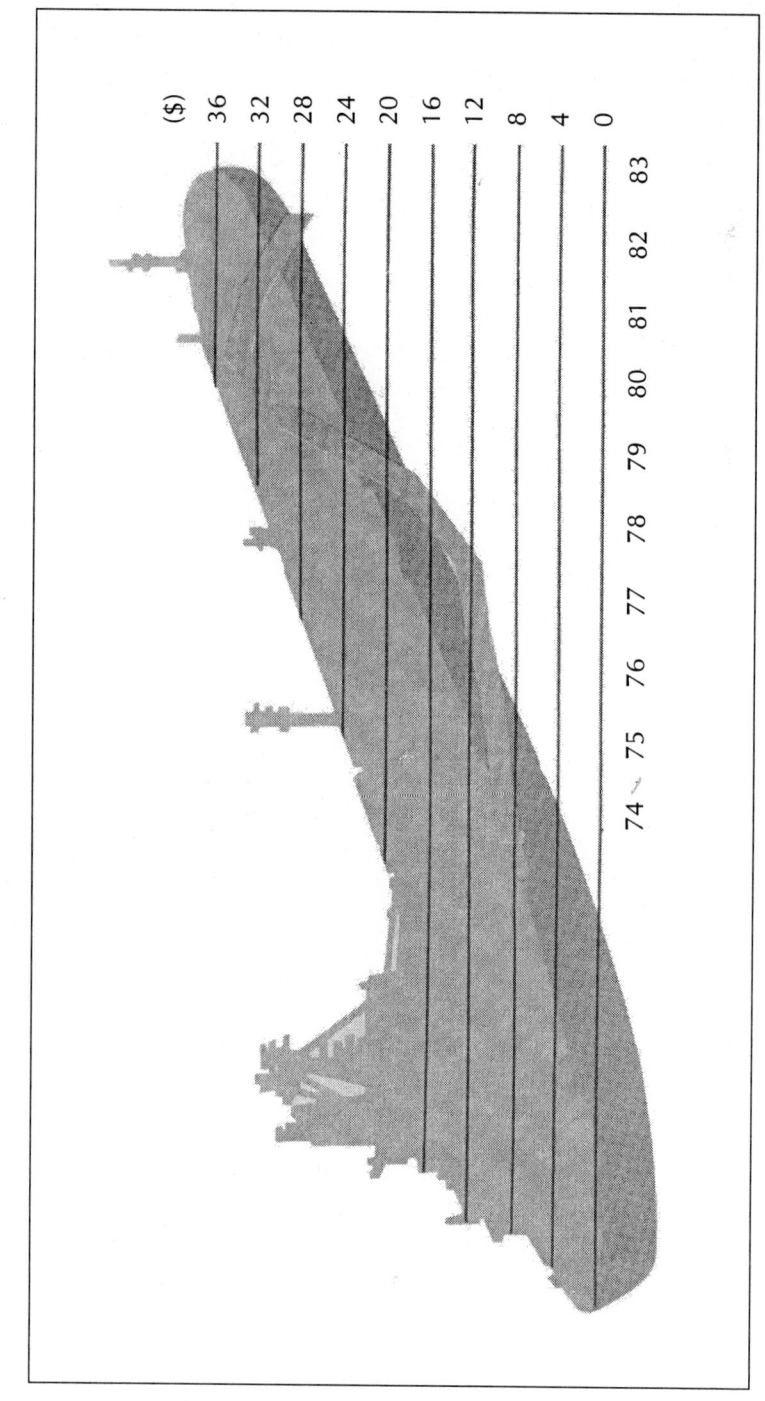

Melchor. Whenever I had a new idea, I consulted them first and, once they approved it, getting the approval of the other directors was easy. Then, whenever we needed an enabling law for a new venture or for financing, we simply drafted a presidential decree in coordination with Solicitor General Estelito Mendoza, and then I brought it personally to the president for approval.

I am not implying that President Marcos simply approved everything we proposed. On the contrary, we had several proposals that he vetoed. You see, the president was very legalistic and, hard as it may be for you to believe, he was conscious about following the law. For example, at the height of geothermal development, I wanted to liberalize the entry of foreign firms and allow them full control of geothermal operations, instead of being limited to partnerships with government or private local corporations. The president disagreed and reminded me that my idea was against the constitution, which expressly stated that the state exercised national patrimony over natural resources.

PNOC's growth was phenomenal in terms of size and profit (see PNOC's Financial Position, 1974-1985 on pages 86 to 88). In fact, PNOC was the only Filipino-owned and -controlled company that landed on *Fortune* magazine's list of top 500 firms outside the United States for four consecutive years, from 1978 to 1981. In 1979, we became the first government-owned corporation to pay out a cash dividend of over Php 190 million to the Philippine government. The company's profitability lay primarily in Petron, which had grabbed more than 40 percent of the market by the early Eighties. This remarkable performance of Petron enhanced our credit standing to the point that we had more credit lines than any other company. As PNOC's Vice President for Finance Edgardo M. del Fonso used to say, "We were very liquid and all the banks looked to Petron/PNOC as a prime credit in terms of profitability and liquidity, and in terms of the stability of the business."

Another feather in our cap was the public recognition given by the World Bank and the Asian Development Bank (ADB) to our efforts. The World Bank extolled the country's national energy program as a "model for Third World countries." Also, in a joint evaluation of the country's energy sector, the World Bank and the ADB concluded that the Philippines was one of the few developing countries that "could hope to become self-sufficient within the next ten years or to reduce their dependence on oil imports significantly." In the same report, the two lending institutions commended the Philippine government for "[having] organized its energy program effectively, adopting sensible and pragmatic policies to guide its own and private sector activities. . .and mobilizing the external financial and other resources for its program."[51] Other external parties also noticed the tremendous gains made by PNOC. The international consulting firm Arthur D. Little in fact made a recommendation to the Petroleum Authority of Thailand to organize itself along the lines of PNOC.

Unfortunately, the PNOC's and the Ministry of Energy's organizational might, financial strength, technical expertise, political clout, and accolades earned from independent observers were not enough for us to surmount the biggest political challenge we faced in our alternative energy program—the opposition to the Bataan Nuclear Power Plant.

---

51. This was cited in my article on energy self-reliance and the debt crisis; see *Energy Forum*, 10.

List of PNOC Subsidiaries

Energy Exploration and Development

PNOC Energy Development Corporation
Incorporated on 5 March 1976, PNOC-EDC undertook exploration for coal, geothermal energy, uranium, and other non-oil resources. It implemented a coal exploration and development program in Uling, Cebu and malangas, Zamboanga del Sur. Major geothermal sites were in Leyte, Southern Negros, Davao and Albay.

PNOC Exploration Corporation
Created on 30 April 1976, it focused on potential onshore oil exploration. Areas of activity were in Cagayan Valley, Central Luzon, Northern Samar, Southern Cebu, and Mindoro.

Malangas Coal Corporation
Incorporated on 22 November 1979, it engaged in the exploration and mining of all kinds of ores, metals, minerals and coal, and all products and by-products thereof.

PNOC Energy Supply Base, Inc.
Established in 24 June 1977, it engaged in warehousing and all business incidentals thereto; provided support and logistics facilities for onshore and offshore exploration activities in petroleum and other related energy fields.

PNOC Geothermal Technology Corporation
Incorporated on 25 February 1980 as a joint-venture company with KRTA of New Zealand, it was involved in providing technological expertise in geothermal development.

PNOC Enercon Corporation
Organized on 26 March 1980, it developed and tested methods and processes to save and efficiently utilize energy from all sources. Also to market, import and export energy-saving mechanisms.

PNOC Energy Drilling, Inc.
Incorporated on 31 July 1980, it explored and drilled for petroleum, oil, gas and other volatile substances, minerals and mineral substances, including soil deposits of every nature and description.

PNOC Alcohol Corporation

Created in 1980, it engaged in the manufacture, production, purchase of alcohol and alcogas or other similar or analogous substances.

PNOC Coal Corporation

Incorporated on 21 January 1981, it engaged in the exploration and mining of coal and setting up and managing a national coal logistics program.

Transport & Logistics

PNOC Tankers Corporation

Incorporated on 19 June 1974, it owned the Philippines' first crude oil tanker, the M/T *Diego Silang*, which had a deadweight tonnage of 96,717 and was capable of carrying 740,000 barrels of crude oil in a single trip.

Petron Tankers Corporation

Incorporated on 13 September 1974, the company owned the M/T *Sultan Kudarat*, an 87,214-tonner capable of carrying 650,000 barrels of crude oil.

Petrophil Tankers Corporation

Incorporated on 13 September 1974, this company owned the third PNOC crude oil tanker, the M/T *Raha Sulayman* with a deadweight tonnage of 130,000.

PNOC Marine Corporation

Incorporated on 13 December 1978, it built, repaired, and converted vessels of various types, as well as handled engineering works and the fabrication of components, modules and prefabricated structures.

PNOC Shipping & Tanker Corporation

Established on 27 December 1978 to handle domestic tanker and bulk oil transport.

PNOC Petroleum Tanker, Inc.

Incorporated on 14 December 1979, it conducted, managed and carried on the business of shipping, tankering, lighterage, barging, towing, transport and shipment of goods, chattels, petroleum and other products of whatever kind and nature. It engaged in marine and maritime commerce in general. This company owned M/T *Raha Sikatuna* (61,230 DWT) and M/T *Petron Gasul* (25,222 DWT).

PNOC Petroleum Carriers Corp.
Incorporated on 10 July 1980, it engaged in shipping, tinkering, lighterage, etc. and other maritime activities, the company owned and managed M/T *Emilio Aguinaldo*, a 98,333-DWT crude oil tanker acquired in June 1982.

PNOC Oil Carriers Inc.
Established on 9 September 1980, it engaged in shipping, tinkering, lighterage, etc. and in the transport and shipment of goods and products in general. The corporation owned M/T *Andres Bonifacio*, which had a deadweight tonnage of 260,000, the largest Philippine flag-carrying vessel.

PNOC Crude Oil Tankers Inc.
Created on 2 September 1981, it was also engaged in shipping, tinkering, lighterage etc. as well as marine and maritime commerce in general, the company owned, operated and managed M/T *Gregorio del Pilar*, which had a deadweight tonnage of 206,971.

Petroleum Refining/Marketing

Bataan Refining Corporation
The largest oil refinery in the country that became wholly owned by PNOC after Mobil sold its shares.

Petrophil Corporation
The country's leading manufacturer and marketer of petroleum products with the "Petron" brand name.

Petron TBA Corporation
Organized on 2 January 1975, it marketed automotive products, including tires, batteries, accessories, and special products.

PNOC Petrochemical Development Corporation
It was organized to establish and operate a proposed petrochemical plant in Bataan to manufacture, process, market, and deal in petrochemicals and petrochemical products.

Filoil Refinery Corporation
Located on a 365-hectare site in Rosario, Cavite, it was mothballed and its main processing facilities were transferred to the Bataan Refinery.

Source: PNOC Annual Report 1982

PNOC's Financial Position, 1974-1985

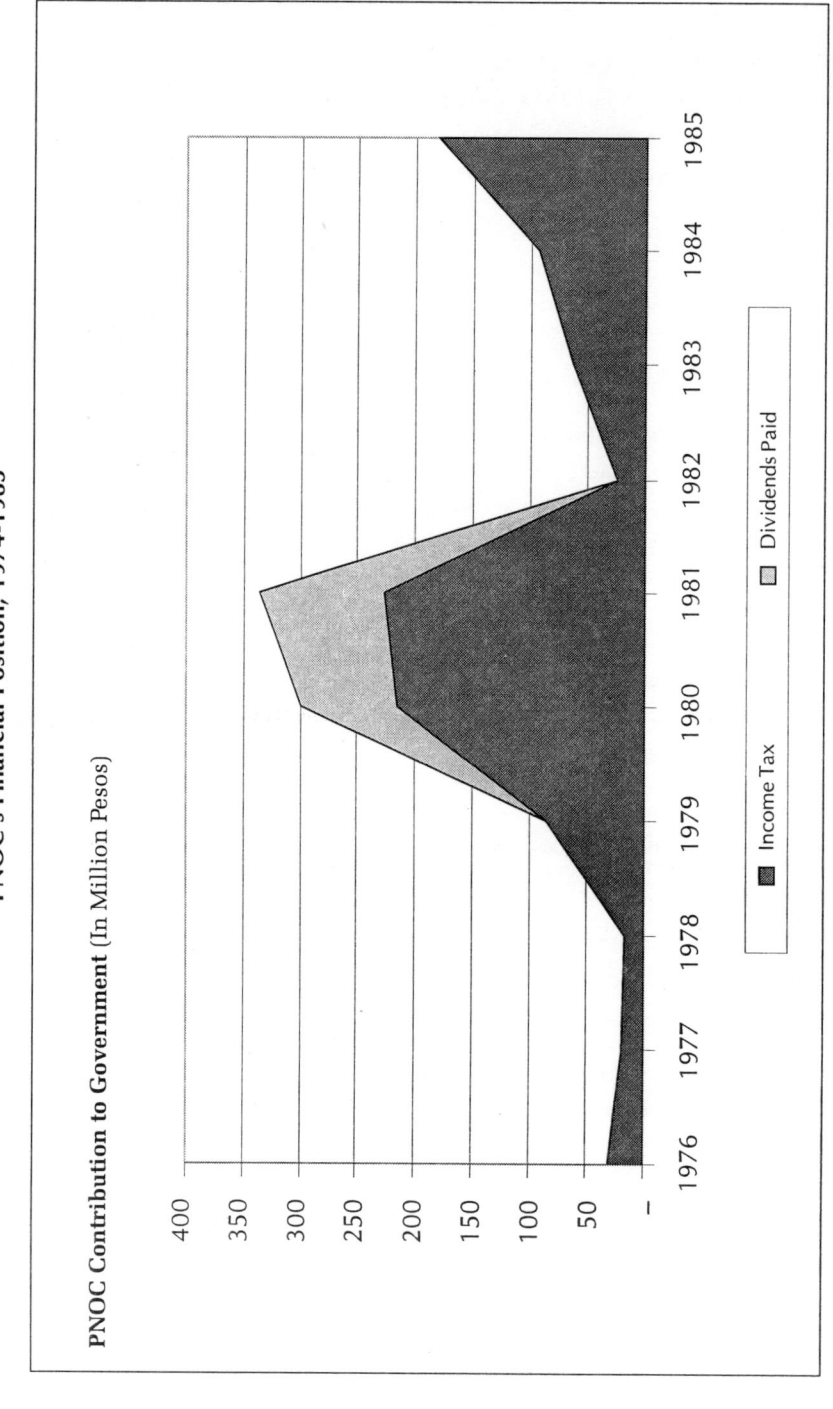

PNOC Contribution to Government (In Million Pesos)

	1976	1977	1978	1979	1980	1981	1982	1983	1984	1985

Legend: ■ Income Tax ▫ Dividends Paid

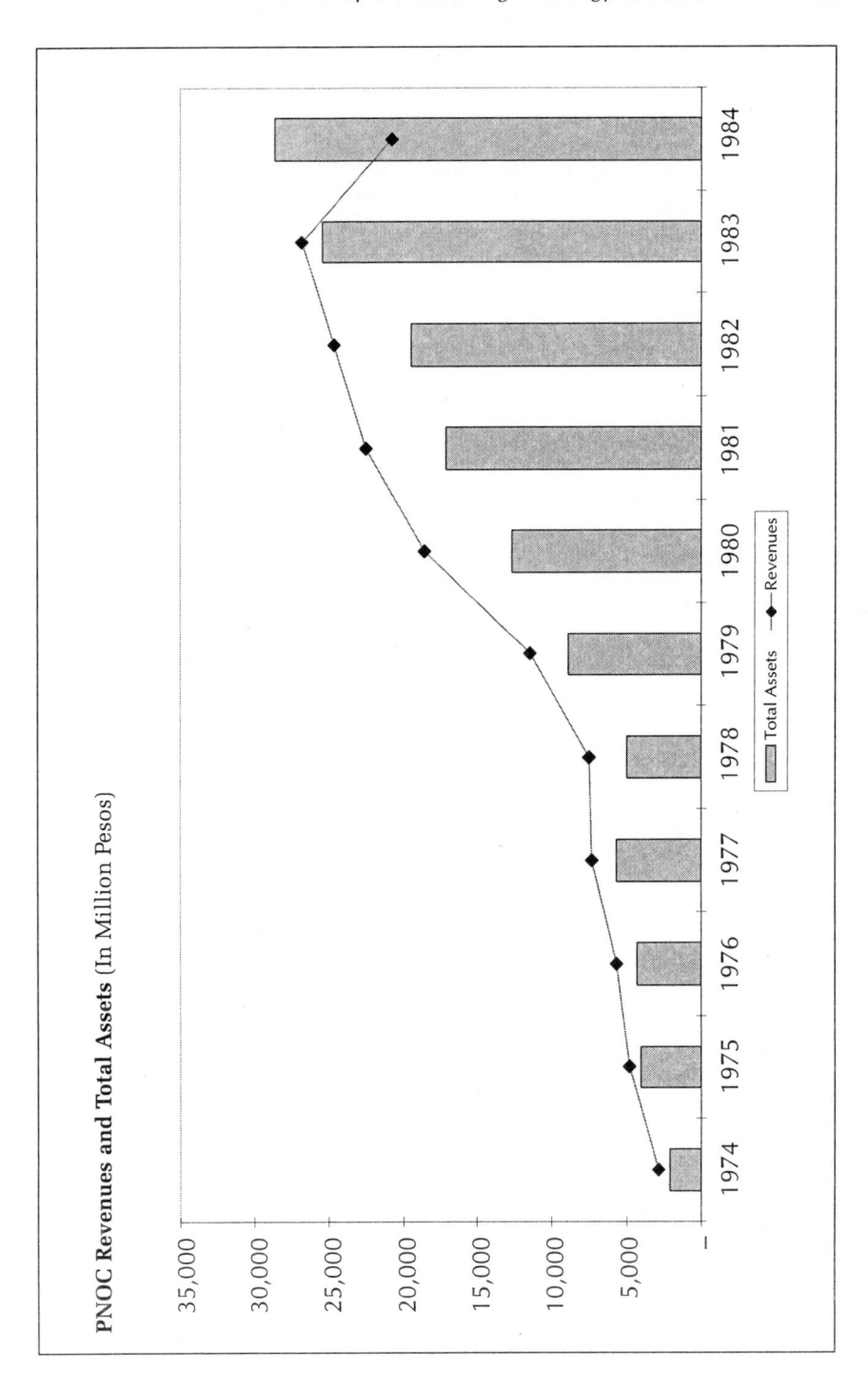

PNOC Revenues and Total Assets (In Million Pesos)

PNOC Net Income (In Million Pesos)

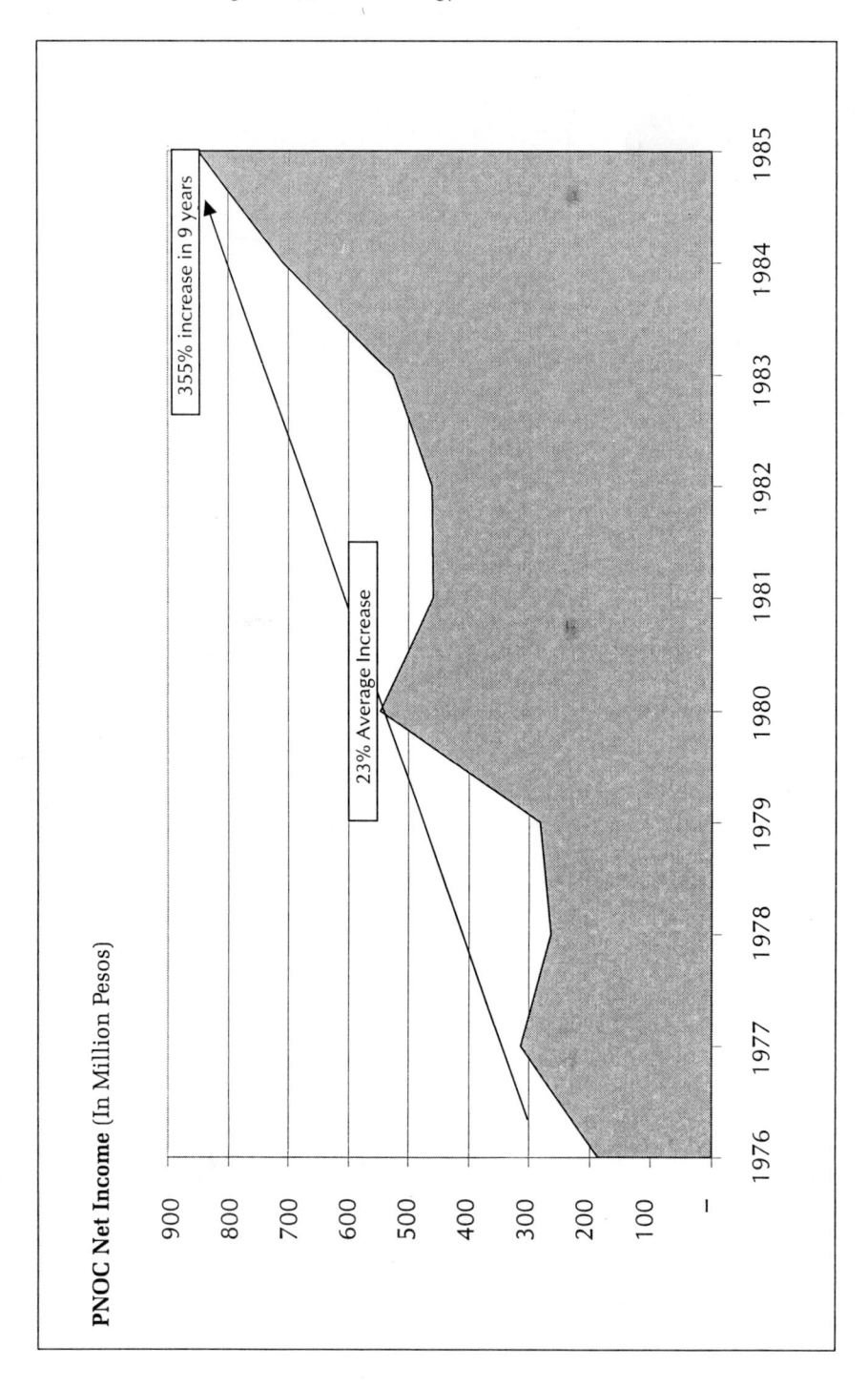

355% increase in 9 years

23% Average Increase

CHAPTER THREE
The TRAGEDY of the BATAAN NUCLEAR POWER PLANT

Nuclear energy is the cleanest form of energy. For a Third World country like the Philippines to operate a technologically advanced system like the nuclear power plant would have been the single biggest achievement of our energy-development plan. The plant could have reduced our dependence on oil, averted the problems we now face over high electricity rates, and provided opportunities for Filipino scientists to learn more about the peaceful uses of nuclear energy. But the anti-Marcos opposition successfully packaged this pioneering project as the ultimate symbol of corruption. As stated by its most prominent opponent, the late Senator Lorenzo M. Tañada, the Bataan nuclear power plant was a "monument to corruption, greed, and folly." Ironically, the icon of the anti-nuclear plant movement was the same man who, as a senator in 1968, ensured the passage of the very law that enabled the Philippine government to venture into nuclear power development. But that is another story.

GZV at the regular TV program "Energy Forum" with Sr. Deputy Minister Gabriel Y. Itchon, concurrently president of the National Power Corporation. (August 24, 1981)

GZV at the signing ceremonies for the supply of uranium for the Bataan Nuclear Plant, with Deputy Prime Minister J. Douglas Anthony of Australia, in the presence of President Ferdinand E. Marcos. (August 25, 1978)

No other endeavor within the Ministry of Energy was more controversial than the nuclear power plant in Bataan. At the time, mere mention of the word "nuclear" connoted not only bad taste but also bad judgment, at least in the eyes of the project's critics. This situation, no doubt, bothered me. All throughout the period that I was in office, I thought that the nuclear plant project was the most challenging and sophisticated energy form that had ever been put before us. Unfortunately, because of misguided (and myopic) politics, "nuclear" literally became a bad word not to be mentioned in decent society.

All of this amazed and annoyed me, especially when opponents of the project made it appear as though we built the nuclear power plant on a whim. What they did not know, or perhaps ignored at their convenience, was the fact that the idea of using nuclear power for Philippine energy requirements has had a long history, dating back to the time of President Ramon Magsaysay. Of course, President Marcos brought this long-germinating idea to its fruition, but he only built on the efforts of his predecessors. His decision to build the plant must be situated within our overall energy situation at the time. Remember, at the height of the energy crisis in 1973-74, all countries were in search of feasible alternatives to crude oil, and nuclear energy was regarded as one of the most feasible substitutes. The Philippine government shared that thinking. In the context of the country's energy plans, the nuclear power plant was expected to provide 16 percent of total electricity requirements for the Luzon Power Grid, as well as to reduce our importation of crude oil, which was the other major source of electricity at the time.

The effort to harness nuclear energy began in 1955, during the Magsaysay administration, when the Philippines signed up to the initiative of the United States to promote the peaceful use of nuclear energy under the auspices of President Eisenhower's "Atoms for Peace Program." Over the next fifteen years or so, countries that were heavily dependent on oil for power generation undertook various studies to determine the feasibility of using nuclear power plants. In the Philippines, the private utility firm Meralco commissioned two such studies: in 1957 when the American firm General Public Utilities Corporation still owned the company, and in 1967 when Meralco was already owned by Mr. Eugenio Lopez Sr. and other Filipino shareholders. Meralco's commercial interest in a nuclear plant possibly influenced our legislators, including Senator Lorenzo Tañada, in enacting Republic Act 5207 *("Atomic Energy Regulatory and Liability Act of 1968")*, which made it *a matter of state policy "to encourage, promote and assist the development and use of atomic energy for all peaceful purposes."*[52] RA 5207 provided the legal framework that enabled the Marcos government to make nuclear energy an important part of our energy development policy, leading to the construction of the country's first nuclear power plant (see *Timeline of Events Leading to the Bataan Nuclear Power Plant* at the end of this chapter).

My involvement with the project officially began on October 6, 1977, when I became chairman ex officio of the National Power Corporation (NPC or Napocor) although, as NPC director in 1976, I already knew about the issues against the nuclear plant that were being raised by those who opposed the project, particularly on expense, safety, and corruption. I will address the last two issues later, but first I want to make some clarifications about the project's status *as I found it*. These clarifications will probably help us better understand the complex issues surrounding not only the project costs, but also the role of Westinghouse in determining these costs.

Negotiations between Napocor and Westinghouse over the procurement of nuclear reactor began sometime in 1974 after the government decided that, based on a comparison of global sales, Westinghouse's pressurized water reactor was a better choice than General Electric's boiling water reactor. But the process of finalizing the contract proved to be an arduous one, given the many substantive disagreements between NPC and Westinghouse representatives over the terms of the

---

52. Republic Act 5207, "Atomic Energy Regulatory and Liability Act of 1968," June 15, 1968.

contract. The higher-ups in Malacañang Palace were becoming displeased. Finally, Executive Secretary Alejandro Melchor (who was also NPC director) intervened. He literally hauled off both parties and locked them up in a Philippine Navy ship, telling them, "You will not get out of this ship until you produce a contract!" The impasse was broken.

National Power formally signed the basic contract with Westinghouse Electric SA, a Swiss subsidiary of Westinghouse, on February 7, 1976. Westinghouse Electric SA subsequently designated Westinghouse International Projects Corporation (WIPCO) as project implementer. In turn, a contract with parent company Westinghouse Electric Corporation was signed by WIPCO to formalize its role in the project. Frankly, the contract between NPC and Westinghouse was not a standard one in the sense that it was not equitable for both buyer and seller, which is the principle upon which most contracts are based. I had misgivings about the contract because some provisions really favored the seller, Westinghouse.

The first thing I did upon becoming NPC chairman was to request Sycip, Gorres and Velayo, which had been the auditing firm of the Philippine National Oil Company (PNOC), to conduct a comprehensive audit of the project for the purpose of determining not only the accomplishments made before my term, but also the tasks that remained to be done. We also conducted a legal audit, wherein I assigned a team of lawyers headed by Atty. Mario T. Meneses (who was then PNOC's legal counsel) to review the contract and other pertinent documents. The audit's general findings showed deficiencies that were detrimental to Napocor's, and ultimately the country's, interests. In the course of the legal audit, we also discovered that even before the contract was signed in 1976, there had been opinions rendered by Solicitor General Estelito Mendoza and then-Assistant Solicitor General Hugo Gutierrez (who eventually became a Justice of the Supreme Court) that pointed out onerous provisions. Yet, notwithstanding such observations, the contract was still approved and executed.

I asked Mario to help me recollect the details of that period when the project was being audited. Among the things that shocked us were the luxurious expenses that could not be justified under government rules, such as the staff housing at the project site that was comparable to the houses in the Makati enclaves. Westinghouse even built a beautiful house on top of a ridge, with a commanding view of the nuclear plant, for the use of one of the subcontractors appointed by the company. Also, the hourly rates charged by Westinghouse's American engineers assigned to the project were really excessive, although these were justified as being standard rates for Westinghouse employees. Then there was the matter of expiring warranty on various machineries that lay idle because of construction delays. These things certainly added to the very expensive nature of the project, but they were trifling compared to the fundamental issues about the contract.

The first question with regard to the nuclear plant was, how come the prices escalated so much? Well, in the original proposal, Westinghouse was limited to the design of the plant and the supply of equipment, while NPC was to take responsibility for actual construction. However, before the contract was finalized for signature, the Philippine government and Westinghouse agreed that the latter would assume overall responsibility instead, including appointing subcontractors and supervising the construction of the plant's building facilities. At the time, too, a decision had been made to transfer the project site from Kabayo Point in Bagac to Napot Point in Morong, Bataan. Note that this was done to answer the concerns regarding the danger of tidal waves as Napot Point is 18 meters above sea level. However, this decision resulted in increased cost to NPC of US$ 24 million, broken down into US$ 8.2 million for equipment modifications and US$ 15.2 million for civil works. Subsequently, when Westinghouse proceeded to design the plant's physical facilities, naturally, the cost estimates had to be increased to cover these additional responsibilities. This was the principal reason for the enlargement of project costs, and then, the subsequent rise in US interest rates to an unprecedented 21 percent per annum caused further escalation in prices.

Besides the costs and questionable expenses, what bothered me and my team was the extent to which Westinghouse exercised full control over the managerial, operational, and financial aspects of the project. Even if this was our first exposure to nuclear power plants and hence the required expertise was completely foreign to us, I was not comfortable that there were a number of unknown factors that only Westinghouse understood. Napocor had no say at all in any major concern, even if only to protect its interest as a buyer. Naturally, we felt that to continue honoring the contract *as formulated* was unconscionable.

Based on the general findings of the audit, therefore, I recommended to President Marcos that we renegotiate the contract with Westinghouse. The president readily gave his permission. No one who was familiar with the project's background missed the irony that, in the first place, the president himself approved the contract. But my sense is that he had no reason to disagree with our findings and he probably thought that if we could still obtain improved terms from Westinghouse, then there was no harm in reopening negotiations. In fact, he immediately constituted a negotiating panel for the Philippine government composed of myself, Finance Secretary Cesar Virata, and Solicitor General Mendoza. I also organized a technical team that would negotiate with Westinghouse on the details of the contract; the technical team was led by Atty. Mario Meneses and the late Josue D. Polintan, a veteran NPC engineer whom I appointed manager of the Nuclear Power Department from 1978 to 1982.

These negotiations with Westinghouse began in late 1977. We had several meetings in Manila and Pittsburgh, Pennsylvania (where Westinghouse's headquarters were located), and two weeks of intensive negotiations in New York. A major issue discussed with Westinghouse was the process of effecting payments, of which I would like to quote Mario's own recollections:

> The contract stipulated the establishment of some kind of a revolving fund, guaranteed by a Letter of Credit, from which Westinghouse drew reimbursements based on their own certification of accomplishments. What we wanted, instead, was for NPC and Westinghouse to jointly review what kind of progress had been achieved, before making progress payments. Under the contract, Westinghouse was pretty much free to decide on how much they would collect and on what kind of progress.

In this setup, as Westinghouse withdrew payments, NPC had to continually replenish the fund. What made matters more burdensome was that, Westinghouse could freely order equipment and other materials from any supplier they chose, and NPC simply had to pay up.

Another unusual provision in the contract is that the applicable law was Pennsylvania law (my recollection is hazy but if it was not Pennsylvania, it would have been some other state of the United States). Normally, when a Philippine company enters into a contract with an American or Japanese firm, for example, there is a provision that in case of dispute and the two parties submit to arbitration, the venue would be in a third country, like Singapore. As a matter of principle, the prospect of arbitration in another country has the advantage of putting pressure on both parties to come to a mutual agreement because, among other things, arbitration can be very expensive for both parties—just think of the travel costs. In this particular case, however, Westinghouse did not have to worry about such mundane things because not only US laws would apply, but specifically the laws of its hometown!

We were asking for a lot of things like a reduction in price, extension of the warranties on the equipment, and even participation by NPC in checking the delivery of various materials to the site—if only to count them! But, as Mario himself recalled for me, "Westinghouse took a very strong position that the contract had been executed, and they would not want to amend any part of it. We wanted them to reimburse the Philippine government for luxurious expenses that could not be justified, but we made no headway. Their basic argument was that the contract was executed freely and voluntarily."

Of course, Westinghouse did grant minimal concessions, like extending the warranty on the machinery, but there was nothing really substantial; we could not even get NPC personnel to check on the company's inventory of project supplies and materials. And it took much bargaining before they agreed to transfer to NPC the rights to that beautiful house on top of the ridge; after all, NPC paid for the house's construction—never mind if National Power had no use for it.

In all these negotiations, I was assured by the president of Westinghouse himself, Gordon Hurlbert, that the contract was typical of what they entered into with other countries. At several meetings in Manila and in Pittsburgh, Hurlbert was adamant in stating that ours was a standard contract: "That's the way other countries have signed, and that's the way our contract will be." He cited similar contracts with South Korea for a 564-MWe nuclear plant named Kori-I, and with Brazil for a 626-MWe Angra-I plant. I asked Hurlbert if he could give me copies of those two contracts but he refused, saying that these were proprietary interests and the two countries could sue Westinghouse if they found out that the latter allowed us to see the contracts. In the same manner, Hurlbert assured me, they would not disclose our contract to any third party.

It became clear to me that I was faced with a given: *The contract had been signed, sealed, and delivered.* Maybe it is fair to assume that Westinghouse was quite deliberate in denying all our requests for substantive amendments to the contract because, after all, they knew we had no other recourse. Besides, if you were in the position of Westinghouse, why would you agree to change the contract when you had it all firmed up to your advantage?

Honestly speaking, even at the time, I was not very confident that we would make any real headway in amending the contract. From what I knew about the history of NPC's negotiations with Westinghouse, the latter obviously enjoyed a tremendous advantage since it had the support of the United States government.

I want to stress that the Philippines could not even have proceeded to negotiate with a supplier of nuclear machinery, had we not first obtained the explicit approval of the US government, as without this approval no one will process the uranium for conversion into nuclear fuel. Actually, the country's choice was limited to only two suppliers, either General Electric or Westinghouse, as these were the only companies capable of converting nuclear fuel for power generation. Westinghouse obviously knew all along that renegotiating the contract was a futile exercise, since the Philippine government (through NPC) had already obtained the permission of its US counterpart for the construction and installation of the nuclear power plant. This was an ace in Westinghouse's sleeves.

I also recognize that what made the nuclear plant project suspect to some quarters was the involvement of Herminio Disini. I believe that the controversy surrounding the nuclear plant stemmed from the fact that, even before construction began, there were already strong accusations of graft and corruption involving Westinghouse and Disini. That the latter was a key figure in the awarding of the contract to Westinghouse opened a Pandora's Box: *Why was Disini involved? How far up in the hierarchy did this go?* I think if Disini did not enter the picture, the nuclear plant would not have generated as much opposition as it did.

When the project came under my administrative supervision, I purposely confined myself to renegotiating the contract and trying to get improved terms for the country. I refused to be drawn into discussing issues of corruption and simply focused on my job. I remember how Sister Christine Tan, a good family friend, kept chiding me for implementing the project. She always asked, "Why are you doing this?"

In spite of questions about the contract, I personally felt that the project should be pursued. As an engineer, I was convinced of the nuclear plant's technical viability and the promise it held for our scientists and engineers in terms of advancing their knowledge of nuclear-based science and technology. As energy minister, I also knew that the nuclear plant was the last piece in the puzzle, so to speak, in our quest for energy self-sufficiency. We urgently needed to diversify our sources of electricity in order to reduce dependence on imported oil and, in the process, ensure uninterrupted supply of electricity for our growing population and industries.

In retrospect, I think what mitigated my misgivings was the fact that a consortium of foreign private banks led by Citibank, N.A. agreed to finance the plant's construction. Honestly, I interpreted the banks' willingness to extend financing as a clear indication that, despite my reservations and despite accusations of graft and corruption, the required investment was justifiable. I did not think that these banks would allow themselves to be party to a project that is riddled with anomalies; instead, they seemed to lend credence to the viability of the project.

What encouraged me to personally commit to the project's completion was my own conviction that all the technical spadework had been done to ensure that we also had the engineering capability to operate our first nuclear power plant. No doubt, the decision to build a nuclear power plant had the benefit of solid research and feasibility studies. The Philippine government, through the Philippine Atomic Energy Commission (PAEC) and NPC, worked with private consulting firms and international agencies like the International Atomic Energy Agency (IAEA) and the United Nations Development Programme in carrying out various types of study to ensure that we would have a viable endeavor. None of these studies pointed to adverse outcomes that the nuclear plant would bring about. We went through the mill, figuratively speaking, and worked with international agencies; everything we did was consistent with international protocol.

I am emphasizing this because, today, the US and the European Union are alarmed over the presumed nuclear capability of Iran and North Korea. In international circles, many assume that Israel has nuclear capabilities, although how it obtained the wherewithal remains a big question. Since the end of the Cold War a number of countries have been said to have obtained access to nuclear technology from the "international black market," whose existence was confirmed in early 2005 when Pakistan's chief nuclear scientist was accused of selling nuclear know-how and technology to Libya. In contrast, the Philippine government intended to procure nuclear reactors commercially, thereby opening the country up to international scrutiny.

Strictly speaking, for a developing country like the Philippines, the use of nuclear energy as an alternative source will not really contribute much to energy self-sufficiency because we will be dependent on another country to process the nuclear plant's uranium fuel. There are only three countries in the world that have the capacity to convert uranium into nuclear fuel—the US, Russia, and France. But despite this limitation, I think the country would have been in an enviable position if we had the nuclear capability.

Looking back, I think our citizens were not comfortable with the fact that we were handling a very delicate and dangerous material. Ironically, and perhaps it is time that I talk about it, we were then the envy of our neighbors. Malaysia was very concerned that we could develop the capability to build nuclear weapons. Indonesian President Suharto was very interested in putting up a nuclear plant; President Marcos sent me to Indonesia to personally brief President Suharto about our project. I even advised him to start with small experimental and research units, similar to what we have at PAEC, rather than build a commercial-scale nuclear power plant at once. Thailand was also curious. If we had pursued the project, we would be the only country in Southeast Asia to possess nuclear capability. Given the vagaries of regional politics and conflicts at the time, the Philippines could have obtained a vital means to enhance its security and strengthen its military position vis-à-vis neighboring countries. We missed a wonderful opportunity. Even back then, we showed our neighbors that we were technologically ahead of them. The fact that the Americans granted us permission to build a nuclear plant under their auspices showed their confidence in our capacity.

Of course, in those days, I did not think of it that way. I have to admit that I also felt like hiding under the table because every time the nuclear plant was mentioned, the accusation of graft leveled against Disini and even President Marcos inevitably came up. That is why I keep saying that the nuclear plant's opponents found a good issue—that it was the symbol of Marcos corruption—behind which they could conceal their fear that Filipinos could not operate the plant. Yet, today, many countries in the world would like to build a nuclear plant to service their energy needs. For instance, China and India are planning a massive expansion of their present nuclear power-generation capacity so that, by 2020, China will enjoy 36,000 megawatts of nuclear power (from today's 9,000 megawatts) and India 20,000 megawatts (from 2,800 megawatts currently).[53]

Our country had the opportunity to develop its capacity for nuclear power generation thirty years ago, and we declined it. Where is our

---

53. "Hitachi, GE, two others ally on nuclear reactors," *The Wall Street Journal*, December 20, 2005.

common sense? Sure, the Philippines was not the only one that mothballed a nearly completed nuclear plant. Puerto Rico and Austria did so after a referendum—the first because the prospect of a nuclear plant accident was deemed too risky considering the island's small population, and the second because buying electricity from the European Grid (which France was largely supplying with nuclear power) was deemed more efficient than generating its own.

This brings me to the three major issues raised by the anti-Marcos opposition against the project: safety, expense, and corruption. On the first, critics raised two concerns—that the nuclear plant lay along an earthquake fault, and that Filipino technicians could not be trusted to operate the plant. On the issue of expense, the opposition harped on the US$ 2.1-billion project cost and even quoted Ramon Ravanzo, the NPC general manager during the negotiations with Westinghouse, who supposedly said, "Marcos has purchased one reactor for the price of two."[54] Finally, the anti-Marcos activists claimed, the massive corruption supposedly involving Disini, Westinghouse, and President Marcos rendered the project illegitimate. For the sake of fair hearing, allow me to discuss each of these concerns.

Safety

Critics kept repeating that the Bataan plant, located at Napot Point in the municipality of Morong, stood near an earthquake fault line and was surrounded by several active and dormant volcanoes, including the biggest in Luzon, Mount Pinatubo. To the project's opponents, the government seemed to be happily inviting an eventual nuclear accident by situating the power plant in an area "with many fault lines." They extended their argument by saying, in effect, that the Philippines could not afford nuclear power plants because the country sits on earthquake faults.

I want to respond to this argument in a commonsensical manner. Japan has more earthquakes than the Philippines, yet it has fifty-four nuclear

---

54. See Dean Jorge Bocobo's article, "PPA the bitter fruit, BNPP the rotten root," which first appeared on the *Philippine Daily Inquirer* in July 22, 2002 and subsequently posted at *http://www.odiousdebts.org/odiousdebts*.

reactors supplying 34 percent of Japanese energy requirements. Although there have been several nuclear plant accidents, including one in August 2004 at Fukui prefecture's Mihama plant that caused the death of four people,[55] none of these accidents was due to earthquake. Elsewhere, in California, earthquakes are an almost yearly occurrence especially in San Francisco and adjacent areas. Yet, the state ranks eighth among the thirty-one states with nuclear capacity and is home to two nuclear power plants, San Onofre and Diablo Canyon. The four nuclear reactors used by the two plants generate a combined net capacity of 4,324 megawatts of electricity and account for 19 percent of California's electricity market.[56]

I am not belittling the concern over earthquake faults, but I think we also need to know the facts before we make any claims. As for expert opinion, I want to emphasize that the IAEA approved Napot Point as a suitable site for the nuclear plant. I also remember that my colleagues at PNOC, the late Dr. Arturo P. Alcaraz (the country's top geologist and Ramon Magsaysay Awardee) and Dr. Arthur "Art" Saldivar-Sali supported the choice of the site. Art Sali used to explain that a structure's proximity to an earthquake fault does not automatically result in destruction per se. I recall he once told me that Napot Point's geological advantage was that its ground is solid as it is underlain by hard bedrock, such that an earthquake would not shake it as much as it would soft or shallow ground. I think Art had a point.

Subsequently, the "proof of the pudding" came when Mount Pinatubo erupted in 1991. That geologic event generated severe earthquake tremors that shook the ground and damaged several buildings, bridges, and property in many provinces in the plains of Central Luzon where the ground is mostly soft, loose soil. In contrast, the nuclear power plant structure, which is closer to Mount Pinatubo than many of the damaged Central Luzon provinces, had no reported damage! Moreover, no geologic faults in Napot Point were reported to have moved, as seen in the apparent absence of

---

55. *BBC News,* "Accident at Japan nuclear plant," August 9, 2004 (*hhttp://news.bbc.co.uk/2/hi/asia-pacific/3547828.stm*).
56. See the United States Energy Information Administration's report on the state nuclear industry (*http://www.eia.doe.gov/cneaf/nuclear/page/at_a_glance/states/statesca.html*).

ground cracks following the eruption of Mount Pinatubo. I think that proves the appropriateness of the site's geology and the nuclear plant's engineering design.

I imagine critics saying that Japan and California are not comparable to the Philippines because they are more technologically advanced. However, to accept that kind of reasoning, to my mind, is to succumb to defeatism—conceding to self-imposed or preconceived limitations without even trying. Look at India, a seemingly Third World country if we focus solely on the massive poverty of its people, but way ahead of many countries in terms of scientific and technological competence. Various Indian governments invested heavily in science and technology, and maximized their large pool of scientists and engineers to develop indigenous nuclear technology. The result—India has built and operated fourteen nuclear power plants with minimal need for foreign expertise. Pakistan is another country worth mentioning. It built its nuclear capability by buying technology from various sources. If these two countries could do it, then why couldn't we? I do not think the Philippines is technologically inferior; it's just that, it seems to me, the people of India and Pakistan have more confidence in their own scientists' and engineers' capabilities.

As for other safety concerns about the Bataan nuclear power plant, we should remember that the National Power Corporation had to obtain prior clearance from the International Atomic Energy Agency and the Philippine Atomic Energy Commission before it could proceed with project implementation. I can understand if people doubted PAEC's independence under a martial law administration, but we still had the IAEA to look over our shoulders; I think that to doubt the IAEA's independence would already be ridiculous. We also had to deal with the US Nuclear Regulatory Commission (formerly Atomic Energy Commission), which monitored the project to ensure that US laws on the export of nuclear technology were not violated.

Speaking of the US, another thing I want to emphasize is the geopolitics behind nuclear power plants. You cannot just buy one off the shelf, so to speak. In those days (and until now), any country that wanted to build a nuclear power plant needed the support of one of the three major nuclear powers in the world—the US, the former Soviet Union (now Russia), and

France—simply because only these countries processed uranium fuel. Two major conditions were imposed on other countries that wanted to build nuclear plants: First, they had to purchase the nuclear fuel from one of these nuclear powers and, second, the spent fuel rods had to be given back to the latter. After all, the spent fuel was the key ingredient in making a nuclear weapon, and the nuclear powers did not want that material to be readily available to just about any country.

Since the Philippines had a special relationship with the US, the latter's support was crucial in securing all the necessary permits from the IAEA, and the most important permit had to do with the nuclear fuel. We had two basic agreements with the US: We would supply them the uranium for conversion into uranium rods, for which we would pay. Then, we would keep the spent fuel rods in a specially designed on-site storage facility, but we could not use them and the US government had the right to get these spent rods any time.

Given the involvement of powerful external parties, the government could not have built the nuclear plant without properly observing the required safety measures, even if it wanted to. I mean, that was the issue against the Soviet Union when the Chernobyl nuclear power plant exploded in 1986—the IAEA member-countries were highly critical of the Soviets' lack of transparency and refusal to allow third-party inspection. The Philippines certainly did not have the political and military strength of the Soviet Union to act in a similar manner.

Besides, if you look at it from President Marcos's perspective, a shoddy project implementation was not in his best interest. After all, one of the IAEA's requirements for countries wanting to construct nuclear power plants is that the government should act as insurer, because no private insurance company in the world would agree to insure a nuclear plant. By the very fact that the government insures the plant, naturally, it will use its best judgment to make sure that the nuclear plant is safe! That was the value of RA 5207 (as amended by Presidential Decree 1484[57])—besides

---

57. President Marcos signed Presidential Decree 1484 on June 11, 1978, amending RA 5207 ("Atomic Energy Regulatory and Liability Act of 1968") in order to add an explicit provision concerning the extent of government indemnification of contractors and suppliers in the event of a nuclear incident.

providing government with the legal and institutional framework for the exercise of regulatory and licensing powers over nuclear facilities and materials, it defined the scope of liability that the government would assume in case of nuclear damage.

Although I did not really bother about it at the time, the growing resistance to martial law made the political environment hostile to the nuclear power plant. But I think certain events in the United States also influenced antinuclear politics here. As early as 1975, antinuke activists in the US had been campaigning actively against their nuclear power plants, which numbered to fifty-six then. The fear of a nuclear catastrophe in any of these plants was raised by consumer activist Ralph Nader and heightened by books like *We Almost Lost Detroit* by John Fuller and *The Prometheus Crisis* by Thomas Scortia and Frank Robinson. Of course, accidents like the one at the Brown Ferry nuclear plant in Alabama in March 1975, fed the protest. At Brown Ferry, a technician accidentally set fire in the plant's electrical system when he used a candle to search for air leaks. There was no meltdown, but the plant's "redundant safety systems" were disabled.[58]

Then the Three Mile Island nuclear accident in Harrisburg, Pennsylvania happened on March 28, 1979. It was the biggest US nuclear-plant mishap thus far. Due to equipment malfunction and human error, one of the two reactors lost its coolant, causing the radioactive fuel to overheat and leading to a partial meltdown.[59] The nuclear damage was relatively minimal, but not the political fallout that hit the US nuclear industry. Orders for new nuclear reactors were cancelled immediately after the accident, many plants already built did not get operating licenses as the Nuclear Regulatory Commission tightened its rules on safety and, generally, there had been no orders for new nuclear plants in the US since 1979. Equally important, the Three Mile Island event strengthened the cause of antinuclear power plant movements.[60] One interesting tidbit is

---

58. *Time Magazine*, "The Great Nuclear Debate," December 8, 1975.
59. *Time Magazine*, "Perhaps the Worst, Not the First," May 12, 1986 (on the Chernobyl nuclear plant accident).
60. See the following stories from *Time Magazine*, "Pulling the Nuclear Plug," February 13, 1984; "Time to Choose," April 29, 1991 (cover story); and "Nuclear Summer," May 28, 2001.

that the accident occurred in the same week that *The China Syndrome*, a Jane Fonda movie about a catastrophic accident in a fictional nuclear power plant in California, was showing.

Three Mile Island obviously provided ammunition to the anti-Marcos opposition. Suddenly, they had a real-life and immediate example of what could happen to the Bataan nuclear plant. President Marcos responded to the accident by promptly ordering the suspension of construction activities and by constituting a commission to investigate the safety of the Bataan nuclear plant. I was aware of the fact that national policy dictated that the plant should not in any way pose a danger to our people. As we explained in the 1985 issue of the *Energy Forum*, there were four occasions when the president demonstrated willingness to shelve the project if the safety of the nuclear power plant was put into question. In recounting these four occasions, let me quote the following excerpts from the said article:[61]

> The first time was in 1976. At that time, barely six months after the signing of the nuclear power plant contract, President Marcos directed National Power Corporation (which will operate the plant) and Westinghouse (which will supply the power plant system) to prove that the nuclear power plant can withstand earthquakes and tsunamis similar to those which had just then devastated Mindanao. Otherwise, he said, he will cancel the construction and supply contract for the project.[62]

As indicated above, this concern was addressed by moving the plant site. The project's resumption was approved only after it was shown that the nuclear plant, to be situated now at Napot Point, could not be damaged by similar earthquakes and tidal waves.

> [T]hen, in February 1978, the Union of Concerned Scientists (UCS) of Cambridge, Massachusetts [one of the groups active in the various anti-nuke movements in the US] sent an unsolicited report to President Marcos alleging deficiencies in the project which, if correct, would pose a danger to the people living near the plant site. In response, the government

---

61. "The Philippine Nuclear Power Plant Project," *Energy Forum* (Makati: Philippine National Oil Company), Volume One, 1985, 35-38.
62. Ibid., 35-36.

commissioned Energy Research Group Inc., a US-based organization of engineers and scientists involved in energy assessment and dissemination of energy information, to make an impartial assessment of the Bataan nuclear power plant.[63]

After reviewing the plant's design and specifications, the Energy Research Group reported that, "We find no basis in the substance of the UCS letter or in the facts surrounding the issues to suggest that the Philippine Nuclear Power Plant (PNPP) cannot be expected to operate safely, reliably, and economically."[64]

> The third time was in 1979 [in the aftermath of Three Mile Island] when President Marcos suspended the construction of the plant and created a commission to investigate the plant's safety features.[65]

I remember when the president confided to me, on the second hole of the Canlubang golf course, that he was thinking of establishing an investigating commission to review the plant's safety aspects, and that he wanted as head of the commission Senator Lorenzo Tañada, a staunch opponent of his administration and the nuclear plant. In the senator's biography, it is said that he declined to head the commission although he agreed to be part of it, and he even had the president appoint fellow anti-Marcos activist Joker Arroyo as member; both men left the commission not long after its formation.[66] The body came to be known as the Puno Commission, since it was headed by Minister Ricardo Puno; its members included retired Court of Appeals Justice Conrado Vasquez.

The Puno Commission submitted its report on November 13, 1979. Its findings essentially supported the continuation of the nuclear power plant, but not until the plant design incorporated additional safety devices. In response, Westinghouse President Gordon Hurlbert wrote to President Marcos on January 17, 1980, to assure him about the safety and reliability

---

63. Ibid., 36.
64. Ibid.
65. Ibid.
66. Agnes G. Bailen, *The Odyssey of Lorenzo M. Tañada* (Quezon City: University of the Philippines Press, 1998), 159-160.

of the plant. "Our reputation and that of the entire nuclear industry," wrote Hurlbert, "is based on the continued safety of our plants. For this reason, PNPP is being designed and will be constructed to be as safe or safer than any nuclear plant in operation in the world today." Still, President Marcos did not lift the suspension.[67]

The Philippine Atomic Energy Commission recommended the lifting of the suspension order on August 15, 1980. In a letter to the president, PAEC said, "[W]ith the regulatory guides, the lessons learned from Three Mile Island, and the IAEA requirements imposed. . .PAEC considers that, under the present state of the technology on nuclear power reactors, the PNPP can be built and operated without causing undue risk to public health and safety."[68]

Thus, project implementation resumed on September 25, 1980 (about fifteen months from the suspension of construction), with additional safety upgrades that cost the Philippine government an additional US$ 100 million, to ensure the prevention of nuclear incidents similar to Three Mile Island's.

> [The fourth occasion where the President showed willingness to shelve the project was] on September 17, 1985 [when] the President directed the Philippine Atomic Energy Commission to extensively investigate and assess the safety of our nuclear power plant before it is given a license to operate.[69]

The actions of President Marcos cited above did not necessarily mean he was affected by the opposition; he knew that all the issues raised against the nuclear power plant were politically motivated, and so he did not really bother with them. But he was sensitive enough not to take the safety aspect for granted.

As I mentioned earlier, the other safety-related issue was whether Filipinos could be trusted to operate the nuclear plant. I tell you, even those sympathetic to the project were a little aghast when they learned

---

67. *Energy Forum*, "The Philippine Nuclear Power Plant Project," 36.
68. Ibid.
69. Ibid.

that the National Power Corporation would operate the plant. People used to say, "What?! You think you can operate that? Do you trust Filipinos to do the job? They would surely forget things and take shortcuts on procedures, and then you'll have accidents!"

Frankly, this issue surprised me. By then I had come to know many talented engineers, geologists, and scientists who were working with PNOC, so the question of whether Filipino technicians could do the job did not even cross my mind. The truth is, Napocor had employed a rigid screening of potential technicians, and the ones chosen were all certified and experienced engineers. As part of their training, we sent forty-six engineers to the Westinghouse training center in the US. At the end of the training, they took the US Nuclear Regulatory Commission test for licensing of plant operators. Of this group, thirty-five passed the exam, two failed and nine went AWOL. While only twenty-four licensed plant operators are required to run the plant, some twelve other engineers were sent for training. Most of this second group passed the licensure exams but some of them stayed behind to work for US nuclear plants; data on this second group are not quite clear. NPC also sent 102 engineers and other professionals to train in various fields, e.g., thirty-six for quality assurance, twenty-nine for maintenance, and nine for health physics.

Expense

Before I address this particular issue, let me just emphasize that nuclear power plants are generally more expensive than conventional power plants because of the nature of the technology and the extra precautionary measures that need to be taken at every phase of construction. As I said earlier, you cannot buy a nuclear plant off the shelf. You have to go through many tedious processes and deal with other parties besides the supplier-company. That is also partly the reason why no nuclear power plant in the world has ever been built on schedule.

As for the Bataan nuclear plant, detractors kept pointing to the US$ 2.1-billion project cost as proof of supposed overpricing. These critics also cited certain statements that I might have made in those days, indicating that I myself found the project a "losing venture" and a "poor investment."[70]

I admit that given the controversy and the mounting opposition to the nuclear plant, there were occasions when I felt the government was better off making use of the resources to put up conventional power plants. Knowing full well that the government was under severe criticism about the project's escalating expenditure, and that the nuclear power plant's costs were really excessive compared to non-nuclear power projects, the seemingly endless cost escalation was very disturbing.

But even back then, other than the luxurious expenses already mentioned, many expenditures were justifiable given the project's many delays, and every safety upgrade incurred additional outlay. We reviewed these costs thoroughly but, really, we could not find an instance where we could say there was overcharging per se. I know that, at first blush, the US$ 2.1-billion project cost is fantastic, but if you review it in its totality, you will find (as we did) that nothing in the contract could make you say in unmistakable terms that Westinghouse was overpricing the nuclear reactor.

What constituted the additional costs? As explained by National Power, initial negotiations only involved the purchase of nuclear power machinery and equipment at a base price of US$ 329 million. Later, due to the complexity of project construction, it was decided that Westinghouse would "also design the nuclear steam supply and the balance of the power plant, construct civil works, erect and test electro-mechanical equipment, manage construction and provide training services."[71] Thus, total cost for Westinghouse's equipment and services amounted to US$ 562 million, at a September 1974 base price. This agreement was contained in the contract signed in February 1976. However, since the delivery of equipment and services would take about seven years, the base price was subject to

---

70. Roberto Verzola, Mae Buenaventura and Edgardo Santoalla, "The Philippine Nuclear Power Plant: Plunder on a Large Scale," in Amado Mendoza Jr. (Ed.), *Debts of Dishonor, Vol. 1* (Quezon City: Philippine Rural Reconstruction Movement, 1991), pp. 51, 66.
71. All information concerning the contract with Westinghouse were taken from the National Power Corporation's *Brief on the Philippine Nuclear Power Plant,* which was issued in 1984.

escalation. The contract already stipulated the formulas for determining the escalation cost, estimated at US$ 115 million until the project's completion. Interest during construction also added a substantial amount to the total cost, as the project took all of seven years to complete.

At the same time, since Napocor had to borrow funds for the project, financing charges were estimated at US$ 266 million. There were actually two sources for the project loan, which originally amounted to US$ 1.109 billion. For the purchase of nuclear technology (equipment, materials, and fuel), NPC obtained a loan of US$ 644.4 million (or 59.44 percent of the total) from the US Eximbank, the only financing institution authorized by the US government to provide assistance in the purchase of nuclear technology. Then, a consortium of private banks led by Citibank, N.A. provided the loan balance of US$ 464.6 million, which would pay for civil works, services, foreign-exchange costs, and others.

Besides paying Westinghouse for equipment and services, Napocor also had to undertake site selection and preparation, fuel procurement and contracting for fabrication and shipment, transmission facilities, insurance etc., all of which were estimated to cost US$ 166 million. Based on the original design, then, the estimated total project cost until completion in 1983 was US$ 1.109 billion, broken down into foreign exchange costs of US$ 833 million, and peso costs worth US$ 266 million.

Substantial construction delays had already translated into huge additional costs. Then, in light of the Puno Commission's recommendation, Westinghouse had to provide safety upgrades. In all, Napocor incurred additional project costs estimated at US$ 844 million, broken down into the following components:

a) Safety upgrades — US$ 100 million
b) Cost escalation — $292 million
c) Financing charges — $373 million
d) Other contract scope changes — $79 million

The estimated project cost had now gone up to US$ 1.95 billion. We must also keep in mind that by the time construction activities resumed in 1980 after a fifteen-month suspension, the Third World was already on the throes of a debt crisis as interest rates soared (reaching 21 percent at one point) and commodity prices fell. Our foreign exchange costs rose from US$ 0.88 billion to US$ 1.51 billion, and since these were financed through foreign loans, interest expense was expected to reach US$ 129 million by 1985. This was equivalent to US$ 355,000 a day, the main additional cost for any delay in project completion. Table 3.1 shows the cost estimates published by NPC in 1984.

As you can see from the estimates, many factors accounted for the outlay of US$ 1.95 billion. Now, I am sure critics will say that given these huge expenses, we should have foregone the nuclear plant's construction. But the Philippine government did not really have that option even if it wanted to change its mind about the project. In the first place, the contract had already been signed, so reneging on it would have had far-reaching consequences for the Philippines. Second, Napocor (as the borrower) had already obtained the first drawdown on the loan from the US Eximbank to pay for the equipment. Worse, at the time when all those delays were happening, Napocor had already taken out part of the loan from the private banks for construction. Hence, with every delay the total loan amount increased such that, by 1984, this had gone up to US$ 1.95 billion (eventually amounting to US$ 2.1 billion by the beginning of 1986). If you were in the government's position, would you actually think that turning back on the project was in the country's best interest?

The next thing that people would probably say, as the late Senator Tañada himself used to say, is that the government should not have accepted a contract with "many onerous provisions" that made the huge project costs possible.

I already recounted our vain efforts to renegotiate the terms of the contract with Westinghouse. Unfortunately, despite my reservations, the contract was valid and binding, and thus had to be implemented accordingly. Worse, as I said earlier, the NPC had already made its first drawdown of the project loan, so even if the government held out and insisted on changing the contract, we still had to pay the bill, so to speak.

Table 3.1
Nuclear Power Plant Original and Revised Cost Estimate
(In US$ million)

	Original	Additional Cost			Estimate
		Safety Upgrade	Delay/ Suspension	New Others	
A. WESTINGHOUSE					
1. Nuclear Plant Equipment	329.4	29.0	26.1	—	384.5
2. Construction & Erection	193.4	31.8	108.9	5.7	339.8
3. Services & Others	39.1	11.0	18.8	23.9	92.8
Sub-Total	561.9	71.8	153.8	29.6	817.1
4. Provision for Escalation	115.2	28.4	94.6[a]	32.1[b]	270.3
Total 677.1		100.2	248.4	61.7	1,087.4
B. NPC Scope					
1. Transmission System	49.6	—	—	(11.1)	38.5[e]
2. Uranium Fuel	48.3	—	—	12.1	60.4[f]
3. Engineering Support & Technical Services	45.3	—	—	10.1	55.4
4. Consulting Services	20.6	—	6.4	14.1	41.1[g]
5. Insurance 2.0	4.7	24.7[c]	31.4		
Sub-Total	165.8	—	11.1	49.9	226.8
6. Financing Charges	266.5	34.5	197.6	141.0[d]	639.6
Total 432.3		34.5	208.7	190.9	866.4
Overall Total 1,109.4		134.7	457.1	252.6	1,953.8
Peso Cost P 1,695.7					P 4,164.6
Foreign Exchange Cost	$ 883.3				$ 1,510.5

NOTES:
  a/ Includes $32.5-M escalation on WIPCO Management extension charges
  b/ Adjustment of $32.1-M on the originally assumed 6% & 10% dollar and peso estimate based on actual trends of 12.0% & 13.6% escalation rates, respectively.
  c/ Includes adjustment on the original estimate which was underestimated, and $4.5-M for nuclear risk insurance
  d/ Includes $109.8-M adjustment of the originally assumed 8% interest charges based on interest rates of actual loan financing (which included floating interest rate)
  e/ Reduction in the original estimate due to change in specs, deleting the 345 Kv line
  f/ Increase of $12.1-M due to provisions for inventory fuel
  g/ Increase of $14.1-M due to expanded man-day budget to cover for additional scope of work, including the pre-service inspection

Now, even assuming that the project did not encounter as many problems as it did and that the construction was completed in a shorter time, the nuclear power plant would still be more expensive than a conventional one. We have to understand that nuclear plant economics is similar to the economics of a hydroelectric plant—both have high initial investment costs but relatively low operating and fuel costs. Some people say that this is an advantage as you lock in your cost in the early stage of the construction. However, in computing for the generation cost, this huge initial outlay is spread over its useful life of, say, thirty years for a nuclear plant and fifty years for a hydro plant, resulting in relatively cheaper cost than an equivalent new oil thermal plant. As a case in point, even if the nuclear plant was completed at a cost of US$ 2 billion in 1985, the comparative generation costs would have been as shown in Table 3.2.

Table 3.2
Comparison of Costs of Different Types of Thermal Plants
(Generation Costs Php/kwh*)

Plant	Fuel	Depreciation	Other OpEx	Int.	Total
Nuclear-US$ 1.1-B	0.17	0.17	0.02	0.35	0.71
Nuclear-US$ 2.0B	0.17	0.30	0.02	0.65	1.14
Oil-existing	0.90	0.04	0.02	0.09	1.05
Oil-new	0.90	0.10	0.02	0.22	1.24
Coal (Calaca-I)	0.46	0.18	0.01	0.14	0.79
Geothermal	0.52	0.09	0.02	0.16	0.79

* Please note that these are all end-1984 costs.

The Bataan nuclear power plant could have been the third-biggest non-oil-based source of electricity (after hydro and geothermal). With an installed capacity of 620 megawatts, the plant needed only about US$ 20 million a year for uranium fuel to supply roughly 16 percent of Luzon's power requirements. It would have displaced an annual rate of as much as 6.5 million barrels of oil, equivalent to about US$ 180 million in yearly savings. Our estimates then showed that, with the plant's full operation, we would have enjoyed annual net fuel savings of about US$ 160 million; at that rate, we could theoretically recover our US$ 1.9-billion investment in the project in less than twelve years. In the face of high crude oil prices at present, I will submit that if the nuclear plant had been allowed to operate, we could have generated savings from oil imports to pay for the nuclear plant's loans. By this time, after twenty years, the nuclear plant would have paid for itself. I maintain that the nuclear power plant's potential benefits still outweigh the costs.

Corruption

The most powerful argument made by anti-Marcos activists against the nuclear power plant was that it became the vehicle for the commitment of massive corruption, which supposedly involved Disini, Westinghouse, and President Marcos. This issue has been the subject of many articles and case studies, and I do not intend to add any more to the exposition of the matter.

Let me just say that whether Westinghouse International Corporation indeed resorted to bribery in exchange for being awarded the contract is a matter of speculation. For whatever it is worth, I wish to mention that Westinghouse cleared itself of such accusation during the investigation conducted by the US Department of Justice under the Foreign Corrupt Practices Act of 1977, which prohibits American companies from making corrupt payments to foreign officials for the purpose of obtaining or keeping business. Under this law, bribing a foreign government entity in exchange for a contract is punishable as a criminal act in the United States. Had the Justice Department found Westinghouse guilty of bribery, the Philippine nuclear plant project would have been exposed to US laws. I mention this investigation of Westinghouse for whatever it may be worth.

But even assuming, for the sake of argument, that corruption accompanied the transactions between the Philippine government and Westinghouse, I still find it strange that the issue of a nuclear power plant is totally disdainful to the Filipinos.

I am not dismissing the gravity of the accusations concerning corruption, but note that these arguments do not delve into the merits of the nuclear plant—whether it was favorable or unfavorable, beneficial, or detrimental to Filipinos and the Philippines. Instead, the nuclear plant was dismissed as the symbol of the Marcos government's corruption and, ergo, bad for the country.

This was my main problem when I assumed responsibility for the project's implementation. No matter how hard we tried to explain the technological soundness of nuclear power plants in general, and the precautionary measures that we were taking in particular, opponents of the nuclear plant kept reducing all arguments to questions of corruption and, as a secondary issue, safety. For all of us at the ministry and Napocor who believed in the plant's technical and economic feasibility, the effort to clarify issues and correct misconceptions seemed useless since our critics would not listen anyway. It was very frustrating, but we did not let up in our effort to communicate with the public. Upon my urging, NPC published a brochure in 1984 to explain the project's background. Through Boo Chanco, who was responsible for the ministry's public relations, we organized media visits to the project site and even sent several media persons to Westinghouse's nuclear power plants in the US so that they could see for themselves how these facilities worked.

One incident I cannot forget was when I invited to Bataan a small group of antinuclear-plant activists who were all women and belonged to the upper crust of Philippine society. Joining us were my good friend and journalist Gabby Mañalac, and my son Jerome who was on vacation in the Philippines. I toured the group around the plant, explaining every aspect of operations and showing how all this contributed to making the plant safe. But rather than give me the benefit of the doubt, the women berated me and raised issues that no longer pertained to the plant's intrinsic merits. Since I already expected that kind of reaction I did not take offense, but my son Jerome did. When we were walking at a slight distance from the women, he asked me why I allowed "these people" to treat me in such an insulting manner. I answered half-jokingly, "Son, this is what you call 'democracy'." I did not realize that Gabby overheard my remark until the following day, when I read his newspaper column. He used my statement to Jerome as the title of his piece, which was about the visit to the nuclear plant.

Frankly, if it was all up to me, I would have proceeded with the plant's operation. There was simply no scientific or technical reason not to do so. I was disturbed when the antinuclear-plant activists brought in Robert Pollard of the Union of Concerned Scientists. He and his group were professional critics—they were part of all those antinuke activities in the US, so how could Pollard be expected to give an objective assessment of the Bataan nuclear plant? Yet, the anti-Marcos, antinuclear-plant activists hung on to his statements as if his was the last word on nuclear power plants!

Do not get me wrong. Nuclear technology is a dangerous technology; handling it is no child's play, and you need to build the competence and discipline to manage it properly. But this is not the same as painting a wholly black picture of nuclear power plants. Although accidents have happened at nuclear power plants, these were not of the magnitude that justifies the vilification of things nuclear. For all its seriousness, the Three Mile Island incident in Pennsylvania did not cause the loss of a single human life. In the case of the Chernobyl facility in Ukraine, yes, the meltdown and explosion of the reactor had likely led to many deaths (possibly thousands of lives lost, as the Western press speculated at the time), but the accident was also due to the fact that the Chernobyl plant did not have a containment vessel and therefore did not meet IAEA standards. The kind of nuclear reactor used was already very old even back in April 1986, when the explosion happened, and the plant operators did not observe standard procedures. In other words, Chernobyl was an accident waiting to happen.

To my mind, the more relevant issue pertaining to a nuclear plant's safety is the disposal of nuclear waste, which has not been addressed internationally even as of this writing. There have been many attempts in seeking a common solution, but geopolitical considerations often affected the outcome of these efforts. The nuclear superpowers (US, Russia, and France), who have the most inventory of nuclear waste, have their own interests so that any common international solution must also tend favorably to theirs. Needless to say, this is a very difficult objective to meet. At this point, each individual country that operates a nuclear power plant disposes the nuclear waste in a glass-enclosed casing, which is then buried underground in on-site storage facilities. At the same time, no country can just export its nuclear waste—a valuable ingredient in making a nuclear weapon—without going through the intense scrutiny of the IAEA and the nuclear powers. But opponents of the Bataan nuclear plant did not bother much with waste-fuel storage, and simply hammered away at the supposed illegitimacy of the project.

Unfortunately for the nuclear plant, external circumstances conspired against its operation. The assassination of Ninoy Aquino in August 1983 placed the president on the defensive, as the anti-Marcos opposition grew in strength and intensity. His position was also made vulnerable by the fact that because of the international debt crisis, worsened by capital flight after the Aquino assassination, the Philippine economy was under heavy strain. Together with endless questions about his health, I think the president felt that the only way he could settle the score was to call for snap elections, which he announced in November 1985. Although we did not realize it at the time, President Marcos's position had significantly weakened not only domestically, but also in relation to the United States.

In December 1985, when we had already obtained permission from both the IAEA and PAEC to start up the nuclear plant (which was already 97 percent complete), Stephen Bosworth, then US ambassador to the Philippines, called on me and requested that the US government be allowed to send a team that would help us evaluate the plant's readiness for operation. I advised Bosworth that I would clear the matter with the president, which I did. President Marcos granted the request in order to have one more expert assessment on whether the plant met all the safety requirements.

The American mission was composed of representatives from the departments of state and defense, the Nuclear Regulatory Commission, the Los Alamos Proving Grounds, and the Savannah Georgia Institute (an entity engaged in nuclear activities). It took them only a week to complete their work. In his letter to the Philippine government, Bosworth relayed the mission's recommendation that we do not operate the nuclear plant until three conditions have been met:

1) That we build more ingress/egress roads to and from the nuclear plant, in order to have improved accessibility especially in cases of emergency;
2) That we should have a built-in capability to notify the head of state instantaneously should a nuclear accident happen; and
3) That we make available more hospital beds within a 100 km.-radius, as part of providing for contingencies in case of an accident.[72]

It is worth pointing out that the three conditions were matters fairly external to the plant, and the mission said nothing about the readiness or acceptability of the equipment in the nuclear plant. I immediately conveyed my observations to President Marcos, who nonetheless advised me not to start up the plant. He said, "Let's wait until after the elections," explaining that he wanted to concentrate on the special presidential elections for February 1986, and that he did not want any more opposition to the nuclear power plant to distract him. He even went to the extent of telling me, "Don't make any move without getting my clearance." So, with a heavy heart, I relayed the president's message to NPC.

At the time, I interpreted the US mission's recommendations as simply a manifestation of their concern about the country's internal security. I thought that perhaps, the growing communist insurgency (which was quite strong in Bataan), the increasingly unfavorable foreign press coverage of the president and his wife, and the uncertainty over political succession compelled the US government to proceed with caution. On hindsight, though, and having seen all the things that happened from December 1985 to February 1986, my sense of that particular period is that the Americans had lost faith in President Marcos, and they could not trust him to have such a powerful weapon in his hands. Had they allowed the nuclear plant to operate, the president (or his successor) would have had the capability to build a

---

72. The letter of Ambassador Bosworth advising the Philippine government about the conclusions of the mission is on file at both the NPC and the Office of the President of the Philippines.

nuclear weapon—which is what the US is currently trying to withhold from North Korea, Iran, and other "rogue countries."

I waited out the presidential campaign frustrated over the seemingly endless obstacles that stood in the project's way. National Power was all set to operate the plant—we already had a complete fuel cycle. Let me tell you how we brought in the fuel cycle. We had to be very cautious. Initially, we thought of chartering a private airplane to bring in the fuel rods; the plane would land at Clark Field since it was near Bataan, and then we would load the uranium rods into trucks. However, we could not get the special chartered plane that we wanted. Finally, the rods were flown in through an ordinary airline, which landed at the Manila International Airport between two and three in the morning. The fuel was immediately loaded in a truck, which drove to Bataan under heavy military escort.

When I look back at those days in December 1985, sometimes I tell myself that maybe I should have tried harder to convince the president not to accommodate the last-minute US mission any longer. What I found onerous about that mission was that their recommendations had nothing to do with the plant—its integrity or its location—but with other considerations that I thought were flimsy. Really, the Americans' message was simple—"Don't operate that plant." Knowing President Marcos, however, I suppose he had already sensed that the US government was apprehensive over the country's political situation, and did not want to pick a fight with the Americans at a time when he was confronting the most serious political challenge to his presidency.

Nevertheless, I was confident (or perhaps wishful) that once the elections were over, there would be no stopping us from operating the nuclear plant. Of course, I expected President Marcos to win. I was confident that he would survive this latest challenge, and I did not worry a bit about whatever political scenarios might unfold after the elections. I did not anticipate that the snap elections would spell the end of the Marcos government. And that included me.

When Corazon "Cory" Aquino became president after a military rebellion, one of her first orders was the mothballing of the nuclear power plant. About four months later, in June 1986, the new government formally decided not to operate the plant, and advised the National Power Corporation "to stop the implementation of the nuclear plant contracts," including "claims for payment relating to the Philippine Nuclear Power Plant No. I (PNPP-I), including foreign loan agreements, by refraining from effecting interests payments under the said loan agreements" beginning July 1, 1986.[73] Then, in October, Mrs. Aquino ordered the transfer to the national government of the nuclear plant's equipment, material, facilities, and records, as well as the balance of NPC's foreign and local loan obligations.[74]

Although the government's basis for assuming Napocor's loans seemed reasonable (i.e., the mothballing of the nuclear power plant deprived the company of potential revenues from which to repay its obligations), I did not expect Mrs. Aquino to honor the nuclear plant's debts after claiming that the project was tainted by fraud and corruption. Despite calls by some of her closest supporters not to pay "fraudulent" debts, Mrs. Aquino chose to play safe and promised the country's creditors that her government would "pay every cent" of the country's foreign debts, including the nuclear plant's. What made the situation more awful was that Mrs. Aquino, in the eyes of the Western world, was hailed as the champion of freedom and democracy just as President Marcos had been vilified as a dictator. She could do no wrong, and she could have used the popularity and legitimacy she enjoyed in order to bargain for debt condonation. Instead, she heeded her financial advisers and passed up a great opportunity for the country to free itself from a heavy debt burden. What a shame!

I was in New York tuned in to the television while she was making that famous speech at the joint session of the United States Congress in September 1986. When she promised that she would pay all of the country's debts down to the last cent, my heart sank. The nuclear plant had been mothballed by then. Why pay for it? She was the darling of the Americans

---

73. Executive Order 98, "Modifying Executive Order 55," (Series of 1986), December 18.
74. Executive Order 55, "Transferring to the National Government the Philippine Nuclear Power Plant I (PNPP-I), its equipment, materials, facilities, records and uranium fuel, providing for the assumption of the remaining foreign loan obligations of the National Power Corporation (NAPOCOR) with foreign lender under the loans contracted by the National Power Corporation and guaranteed by the Republic of the Philippines and of the Peso obligations incurred to finance the construction of the said nuclear plant by the National Government, and for other purposes," (Series of 1986), October 1.

and she could have used that to her advantage. I thought, if she found the nuclear plant so abominable and the transactions related to it wrapped up in so much corruption, then why pay for it? Maybe she lacked the experience, and maybe she trusted her advisers too much. I think that particular decision demonstrated, at the very least, her inexperience in governance.

This decision to pay the nuclear plant's debts had far-reaching consequences. It has been said that the government paid out US$ 460 million (Php 9.5 billion at the exchange rate of US$ 1 = Php 21), or between US$ 300,000 and US$ 357,000 a day in debt service, from 1987 to 1989.[75] To think that these payments were made at a time when the economy was experiencing minimal to negative growth.

In 2004, after President Gloria Macapagal Arroyo had declared that the country was in a state of fiscal crisis, we had supposedly paid out the equivalent of US$ 155,000 a day. We will continue to pay for the nuclear plant's bills until 2018. The Bureau of Treasury reported that interest and principal payments are expected to reach Php 2.1 billion in 2005, or Php 5.8 million a day. We have supposedly paid out Php 61 billion over the past twenty years, but we still need to pay Php 6.3 billion in the next thirteen years.[76]

Both the Aquino and Ramos governments tried to reach an out-of-court settlement with Westinghouse, with a deal finalized on October 13, 1995. The settlement called for Westinghouse to pay the Philippine government US$ 100 million, made up of US$ 40 million in cash and two 160-megawatt combustion engines worth US$ 30 million each. In return, the Ramos administration agreed to drop all legal cases against Westinghouse as well as Burns and Roe, and allowed the two companies to resume business operations in the country's power sector (the Aquino government had previously blacklisted these companies).[77]

---

75. Freedom from Debt Coalition, "The Story Thus Far—The Bataan Nuclear Power Plant (BNPP): Odious Legacy of Power and Profit" (Quezon City: Freedom from Debt Coalition, 2004), 13.
76. "Pinoys paying P5.8 million a day for Bataan nuclear power plant," *The Philippine Star*, May 14, 2005.
77. Freedom from Debt Coalition, "The Story Thus Far," 18.

If Mrs. Aquino wanted to scrap the plant, as in fact she did, then there is nothing anyone can do about it. But since, presumably, she believed that the project was nothing but fraud and corruption, and considering her international standing at that moment in time, she could have made representations with the Americans for the Philippines not to pay the nuclear plant's debts anymore, or at least to get a settlement right there and then, because US$ 2 billion is too much to pay for nothing. Considering the rousing welcome she got from the US Congress then, I do not think any legislator would have disappointed her had she asked for debt relief. Why make the country pay, and especially after the "ravages" of the Marcos legacy?

The nuclear power plant had gained such international notoriety that it has been constantly cited as an example of fraudulent loans that deserve to be repudiated. In her book, antidebt advocate Noreena Hertz even had an extended discussion on the project:[78]

> [T]ake the Bataan Nuclear Power Plant in the Philippines, built in 1976 for over US$ 2 billion with loans largely provided by the United States' Ex-Im. The largest and most expensive construction project ever undertaken in that country, the loans taken out to build it are still costing the Philippines US$ 170,000 a day to service and will continue to do so until 2018. This in a country in which GDP per capita is US$ 4,000, 40 percent of the population live below the poverty line and annual per capita expenditure on health is only US$ 30. And all this expense for a plant that never worked. "Filipinos have not benefited from a single watt of electricity," said the [former] Philippine national treasurer Leonor Briones. Thankfully not, because the plant's design was based on an old two-loop model that had no safety record of any sort, and because the plant lies along earthquake fault lines at the foot of a volcano.

Another thing I want to take issue with is the settlement with Westinghouse. Look at the way it handled the problem. For all the troubles that the nuclear power plant supposedly caused the country, Westinghouse

---

78. Noreena Hertz, *The Debt Threat* (New York: Harper Business, 2004), 50.

got away for a measly US$ 100 million, including two units of combustion engines whose purpose we are not even aware of. Really, the only settlement the Ramos administration can claim is the cash payment of US$ 40 million. *Eh, barya lang 'yon!* (That's small change!) Meanwhile, Filipinos will be paying through their noses until 2018 for a power plant that stands not only as a "monument to corruption," as claimed by the opponents of President Marcos, but also to careless decision making by his successors.

I am fully aware that the Marcos government will always be blamed for having incurred those loans for the nuclear plant, but Mrs. Aquino cannot escape responsibility for the fact that she agreed to repay these loans despite having mothballed the plant. It saddens me to know that, for all the vilification heaped on the project, Filipinos were nonetheless forced to pay a heavy price in exchange for supposedly maintaining the country's good standing with our international creditors—the same banks and financing institutions that encouraged us to sink deep in the quagmire of questionable debts. It is ironic that amidst accusations that international lenders "propped up" the Marcos government with "fraudulent loans" which did not benefit the Filipino people, his successors chose to legitimize the same. The present government continues to prioritize paying off loans such as the nuclear power plant's, which makes our creditors happy, even as nearly seven million Filipino families live on less than US$ 1 a day.

I want to emphasize that, to this day, I believe in the nuclear plant's viability and necessity. But I also believe that if our people cannot benefit from it, then there is nothing honorable in repaying loans that only make heavier the ordinary Filipino's daily burden.

Timeline of Events Leading to the Bataan Nuclear Power Plant

1955	On July 27, the Philippines committed itself to the peaceful use of atomic energy by signing an agreement with the United States under the "Atoms for Peace Program," which was first proposed by US President Dwight D. Eisenhower in an address before the United Nations General Assembly.
	The RP-US agreement on atoms for peace was the first bilateral agreement signed under this program. Raul Leuterio, charge d'affaires ad interim of the Philippines, signed for the Philippine government. Signatories for the US government were Walter Robertson, Assistant Secretary of State for Far Eastern Affairs, and Lewis Strauss, chairman of the US Atomic Energy Commission.
1956	On October 26, eighty-two nations, including the Philippines, gathered for a United Nations meeting in Geneva and established the International Atomic Energy Agency (IAEA). Its main objective was to encourage the development and practical application of atomic energy for peaceful purposes, including the production of electric power, with due considerations for the needs of the underdeveloped areas of the world.
1957	The IAEA statute came into force on July 29.
	Meralco, then a wholly owned subsidiary of a US-based company, commissioned the American consulting firm Gilbert Associates to make a preliminary study on the feasibility of a nuclear power plant. This study concluded that it was not yet time to undertake the project, since the local market was too small for the plant's capacity.
1958	On May 8, during the administration of President Carlos P. Garcia, the Philippine Senate ratified the IAEA statute. The statute then entered into force for the Philippines on September 2, upon deposit of the instrument of ratification with the United Nations.
	On June 13, Congress approved Republic Act 2067 ("Philippine Science Act of 1958"), which created the Philippine Atomic Energy Commission (PAEC) and empowered it to conduct or promote research and development of, among others, processes, materials and devices used in the production of atomic energy. Congressional approval was based on a report of the Bicameral Conference Committee composed of Senators Arturo Tolentino, Emmanuel Pelaez and Lorenzo Tañada, and Congressmen Salvador R. Encinas, Canuto M. S. Enerio, and Angel B. Fernandez.

1960	In January, the Philippine government requested the IAEA in Vienna for assistance in undertaking a survey of the prospects for nuclear power in the country over the next decade. Several months later, in October, the IAEA mission began its survey.
1961	In August, the IAEA submitted its survey report under the title "Prospects of Nuclear Power in the Philippines" to the Department of Foreign Affairs. The report concluded that, given the general lack of fossil fuels in the country and limited economic hydro potential on Luzon Island, a relatively large nuclear plant may be able to compete favorably with fuel oil-based thermal stations by the end of the decade. The mission also concluded that nuclear energy could play an economic role in the country's power system.
1962	In December, during the administration of President Diosdado Macapagal and through the initiative of PAEC, the Philippine government requested assistance from the UN Special Fund in determining the technical and economic feasibility of using nuclear power to supplement available conventional energy requirements, towards meeting Luzon's projected power requirements.
1963-65	On June 28, the UN Special Fund approved the request of the Philippine government, with the IAEA subsequently appointed as executing agency for the preinvestment study. The UN Special Fund contributed roughly US$ 439,000 for the study, and the government about US$ 262,000 in counterpart funds and services. The study, begun in 1964, was completed in 1965. It recommended that the Philippines seriously consider the use of nuclear plants in the Luzon Grid by the early Seventies, and that legislation for the regulation of nuclear power and third-party liability for nuclear damage be enacted.
1967	Meralco, now owned and controlled by Filipino stockholders led by Eugenio Lopez Sr., invited bids for the steam supply of a 300- to 500-megawatt nuclear power plant, which was expected to be completed in early 1975. Subsequently, however, Meralco concluded that its Metro Manila market could not support an economic-sized nuclear power plant and that it should wait until the late Seventies or early Eighties, when the market would have expanded.
1968	On May 8, the two Legislative Houses approved the report of the Bicameral Conference Committee on a proposed atomic regulatory and liability statute. The Senate was represented in the Committee by Senators Helena Benitez,

Lorenzo Tañada, and Jovito Salonga, and the House of Representatives by Congressmen Frisco San Juan, Gregorio Murillo, and Amado Arrieta.

On June 15, Republic Act 5207 ("Atomic Energy Regulatory and Liability Act of 1968") was enacted. Its Declaration of Policy (Sec. 2) stated, "It is hereby declared to be the policy of the Philippine Government to encourage, promote and assist the development and use of atomic energy for all peaceful purposes, as a means to improve the health and the safety of workers and of the general public, and to protect against the use of such facilities and materials for unauthorized purposes."

Republic Act 5207 likewise empowered PAEC to, among others, issue license for the construction, possession, or operation of any atomic energy facility. It was under this law that PAEC issued the license for the Bataan nuclear power plant. Without this or any similar law, we could not have constructed our nuclear power plant.

On June 13, or two days before the law's enactment, the Philippines and the US signed a new Agreement for Cooperation that called for an exchange of unclassified information on the application of atomic energy for peaceful uses, including information on power reactors and the use and safety of radioactive isotopes (uranium). Signatories were Salvador P. Lopez for the Philippines, and Robert W. Barnett and Glen T. Seaborg for the US. This new agreement, which was signed in Washington, D.C., superseded the Atoms for Peace Agreement signed on July 27, 1955, by both governments.

On July 15, in Vienna, the Philippines, the US and IAEA entered into an agreement for the application of IAEA safeguards to guarantee that the equipment, devices, and materials supplied by the parties under the Agreement for Cooperation shall not be used for any military purpose. The treaty, which came into force on July 19, was signed by Salvador P. Lopez for the Philippines, John Hall for IAEA, and Jack Vanderryn for the US.

1971 On June 23, President Ferdinand Marcos set up a Coordinating Committee for Nuclear Power Study under Administrative Order 293 to prepare the groundwork for a new and updated feasibility study. A subcommittee was subsequently formed on July 27 to formulate the criteria for site selection, among other functions.

On September 10, Republic Act 6395 was enacted, revising the charter of the National Power Corporation (with Manuel Barreto as chairman) and authorizing NPC to construct, operate and maintain power plants for the production of electricity from nuclear, geothermal, and other sources.

In November, the IAEA initiated a market survey for nuclear power in twenty-three developing countries in the Far East, including the Philippines (where the survey took place from October 23 to November 17, 1972).

1972 The Coordinating Committee for Nuclear Power Study, with support from the United Nations Development Programme (UNDP), conducted a second feasibility study for a nuclear power plant in Luzon. There were four phases of the study: (A) site selection; (B) technical and economic study (C) organizational and financial study; and, (D) Investment report. Local experts took charge of the first phase, while American and Swiss engineering consultancy firms carried out the other three phases of the study.

A. In January 1972, the Nuclear Power Site Sub-Committee, composed of engineers from PAEC, NPC, and Meralco, conducted a survey of possible sites. Based on considerations of topography, geology, faulting, hydrography and land use, five sites were ranked in terms of suitability:

1. Limay, Bataan
2. Bagac, Bataan
3. San Juan, Batangas
4. 1-1/2 kilometers south of Padre Burgos, Poblacion, Quezon
5. Ternate, Cavite

Ultimately, the choice was narrowed down to between Bagac and Limay, in Bataan.

B. In February 1973, a technical and economic study was jointly undertaken by Electrowatt Engineering Services Ltd. of Zurich and Lundy Engineers of Chicago. They also conducted an organizational study of NPC and Meralco. It was anticipated that both utilities would be able to satisfactorily absorb the costs of the nuclear project.

C. In April 1973, the Swiss and American firms carried out the last phase of the survey. In the investment report, they considered the benefits that would result from the savings in operating costs of a nuclear power plant, and the additional savings in foreign currency that would otherwise go to oil importation for the conventional power station. With the economic advantage established, the feasibility study gave positive recommendations for the construction of a nuclear power plant in the Philippines.

D. The feasibility study also recommended that a government body be immediately entrusted with setting up the requisite organization for implementing the project, starting with detailed site surveys, initiation of licensing and safety procedures, and recruitment and training of staff.

Accordingly, the National Power Corporation began setting up the initial organization and studies for project implementation.

1973 In August, NPC engaged the services of EBASCO Overseas Corporation of New York to assist in the evaluation and selection of the best site for the nuclear plant. The site selection study, completed in January 1976, recommended Napot Point in Morong, Bataan.

1974 NPC held initial discussions with General Electric (manufacturer of Boiling Water Reactors or BWR) and Westinghouse (manufacturer of Pressurized Water Reactors or PWR). The choice of the type of nuclear reactor to adopt for Philippine use was based on the clear preference of electric utilities in the West for pressurized water reactors. Westinghouse had sold 36,392 PWRs compared to General Electric's 11,095 BWRs in 1973; from 1971 to 1973, more PWRs had been contracted worldwide—seventy-eight as against thirty BWRs. Given the market's apparent confidence in PWRs, the government chose Westinghouse as contractor and supplier for our first nuclear power plant. Negotiations between NPC and Westinghouse were held over a period of twenty-two months, from June 1974 to January 1976.

1976 On February 7, the contract for the nuclear power plant was signed by NPC General Manager Conrado del Rosario and Westinghouse Electric Corporation Manager for Business Development Thomas K. Keogh. The contract, which took effect in September 1976, provided for the power plant's completion in mid-1983.

On October 21, NPC filed an application for a provisional permit with PAEC to start groundwork preparation for the nuclear power plant site in Bataan.

On December 20, PAEC (headed by Dr. Librado Ibe, a nuclear scientist) granted NPC a license for groundwork preparation for the nuclear power plant.

1977 On July 12, NPC filed an application for a construction permit with PAEC. About three months later, on October 3, PAEC granted NPC its first limited work authority for the nuclear power plant.

	On October 6, the Ministry of Energy was created under PD 1206 and Mr. Geronimo Z. Velasco was appointed its minister. The NPC became an attached agency under the ministry, hence Minister Velasco assumed the company's chairmanship on an ex-officio basis. It was also during this time that Mr. Gabriel Y. Itchon was appointed deputy minister of energy.
1978	On April 24, Mr. Itchon was appointed president of NPC.
	On June 11, President Marcos signed PD 1484, amending RA 5207 with the addition of, among others, an explicit provision regarding the extent of government indemnification of contractors and suppliers in the event of a nuclear incident.

Source: National Power Corporation, *Brief on the Philippine Nuclear Power Plant*, 1984.

CHAPTER FOUR
PRIVATIZING POWER GENERATION
The Tale of NAPOCOR

When President Marcos designated the National Power Corporation (Napocor or NPC) as the sole body authorized to generate electricity for distribution throughout the country, the government underscored the importance of "total electrification" to our national development goals. The president's opponents, however, interpreted this move as another devious attempt to take away business from his archrival Eugenio Lopez Sr., whose company, Meralco, had been involved in power generation and distribution. The Aquino government, in turn, ended Napocor's monopoly of power generation and "returned" Meralco to the Lopezes. Twenty years since President Marcos's overthrow, the National Power Corporation has degenerated from a flagship government corporation to a bankrupt and heavily indebted one—no thanks to politicians and past appointees whose lack of appreciation for and understanding of Napocor's role contributed to this sorry state of affairs.

4

Inauguration of M/T Diego Silang oil tanker: (l-r) GZV, President Marcos, Irene Marcos and Mrs. Erlinda Velasco. (January 12, 1975)

One of the eleven PNOC drilling rigs initially assembled to commence drilling of geothermal wells. (May 1976)

I reached another high point in my government career on October 6, 1977, when President Marcos issued PD 1206,[79] creating the Department of Energy and appointing me as its secretary (the shift to the parliamentary system in 1981 made me a minister). It was, you might say, a logical development in my new profession as a public servant. Our various projects at Philippine National Oil Company (PNOC) helped to build a respectable energy infrastructure within a short period and created a reliable pool of experts that could run the country's energy programs.

I was confident in taking on the position of energy minister partly because PNOC had prepared me for it, and partly because I was able to tap the services of key PNOC executives who constituted the ministry's core. For instance, PNOC Senior Vice President Antonio V. del Rosario became concurrent deputy minister. Even I remained chairman, president and CEO of PNOC. This probably sounds anomalous to present observers of government but, in my time, key public officials could hold multiple appointments. Frankly speaking, if President Marcos did not allow me to bring in the core of PNOC, I would have had difficulty organizing the ministry and coordinating the efforts of the two organizations. On the other hand, had these executives been compelled to give up their PNOC posts upon appointment to the ministry, the company would have suffered at a time when it was undertaking many pioneering projects.

As minister of energy, I became responsible for all government entities involved in the development, generation, distribution, and regulation of energy resources. Outside of the PNOC, the biggest organization that came under my jurisdiction was the National Power Corporation (Napocor or NPC). I was already quite familiar with the company's operations by then because President Marcos appointed me as NPC director in January 1976 in order to coordinate PNOC's supply of fuel oil to Napocor, which

---

79. Presidential Decree (PD) 1206, "Creating the Department of Energy," October 6, 1977.

happened to be the single largest user of oil in the country. With the creation of the Ministry in October 1977, Napocor became an attached agency and I assumed the post of chairman ex officio. It was during this time that I came to work with Gabriel "Gabby" Y. Itchon, who became deputy minister. Six months later, in April 1978, the president appointed him Napocor president.

The National Power Corporation was created on November 3, 1936, through Commonwealth Act 120. Efforts to establish such an organization had actually been under way a few years earlier, through the tireless lobbying of the late Felimon C. Rodriguez. An engineer with the Bureau of Public Works, he was sent to the US as a government scholar in 1930 to specialize in irrigation. After going on observation tours of several hydroelectric power plants in the US, Rodriguez thought of developing the Philippines' potential for hydroelectricity. Upon his return, he lobbied several legislators to enact a law creating the National Power Corporation for the said purpose, but the American Governor-General vetoed the bill twice due to opposition from several American bureaucrats and US-owned companies, particularly the Manila Electrical Railway and Light Company (Meralco). Organized before the war by the US-based General Public Utilities Corporation (GPU), Meralco was the country's foremost generator and distributor of electricity. Not surprisingly, it opposed the creation of a government entity for generating hydroelectric power, since the latter would compete directly with the company.[80]

Finally, in 1936, Commonwealth President Manuel L. Quezon was persuaded to sign the third bill into law thanks to the strong support of US Army hydraulic engineers, Capt. Hugh Casey and Capt. Lucius Clay, who had been recommended by Gen. Douglas MacArthur (Quezon's military adviser) to study the feasibility of developing the country's hydropower systems. Commonwealth Act 120 thus established an

---

80. A valuable source for the history of the NPC would be the book by Gabriel Y. Itchon, Perla A. Segovia and Arturo Alcaraz, *A Short Story of the National Power Corporation: 1936-1986* (Quezon City: National Power Corporation, 1986), which was published in celebration of NPC's Golden Jubilee. Aspects of NPC's early history mentioned in this chapter came from this book.

organization within the government that had the capability to tap our water resources for generating electricity. Rodriguez played a crucial role in organizing Napocor; maximizing his network in the bureaucracy, he recruited some of the best Filipino engineers who had been working in other government bureaus.

National Power targeted Caliraya in Lumban, Laguna, as the site of its first hydroelectric project. However, before it could set about this task, Napocor still needed to justify the project's marketability. Using their influence, Casey and Clay convinced the American-owned Meralco to buy power from NPC and distribute this for Manila's consumption. On October 31, 1939, NPC signed a sales contract with Meralco for the delivery of 80 million Kwh of power from Caliraya.[81] Meralco thus became the first and biggest customer of NPC; more important, the contract with Meralco provided the needed justification for NPC to begin construction in December 1939. Financed through a bond issuance worth Php 8.5 milion, the project was completed in 1944.

After the war, National Power rebuilt its material and human resources, and positioned itself for a major role in the country's postwar rehabilitation and development. In 1950, the management—now led officially by Rodriguez as general manager—set up an engineer cadetship program wherein several staff were sent to Meralco for training in various aspects of power generation and transmission. Over the next twenty years, as NPC built more dams and hydroelectric plants, the company attracted competent engineers whose skills were harnessed for the development of our hydro resources. Many of the well-known engineers then worked for Napocor; on the part of the engineers, it was a badge of honor to be working for the organization. One of my colleagues at PNOC, Jose U. Jovellanos, built his career at NPC and came to head the power planning division. He was the company representative in Washington, D.C., during negotiations with the World Bank over a US$ 20-million dollar loan for the Ambuklao hydro project.

---

81. Corazon H. Ignacio (Ed.), *National Power Corporation: 60 Years, Vol 1: 60 Megawatts of Achievement* (Quezon City: National Power Corporation, 1996), 15.

The Fifties witnessed the construction of several hydro projects in various parts of the country, including the Ambuklao and Binga plants in Benguet, and the Maria Cristina System in Lanao del Norte. The latter eventually was integrated into the entire Agus river system, utilizing the Lake Lanao water resources from Lanao del Sur all the way to Lanao del Norte and broken into seven Agus systems. Napocor's hydroelectric plants were the first major non-oil source of electricity. All of these projects were financed through foreign loans and postwar reparation. Napocor had always been dependent on foreign financing largely because of the huge investment requirements of power plants, and also because of the company's limited capitalization. Frankly, the huge capital requirements made it impossible for the government to even aspire to finance these requirements through equity contribution. The government could use up all its resources and these would still not be enough to cover NPC's needs.

During his incumbency, President Ramon Magsaysay ordered National Power to extend its electrification services to the rural areas. The company began selling power to other distribution companies besides Meralco. The latter, meanwhile, continued to expand its generating capacity by building additional power plants that used bunker oil for the generation of electricity.

As the country picked itself up from the devastation wrought by the Second World War, the consumption of electricity within Manila and eventually the Greater Manila area grew by leaps and bounds, essentially increasing by as much as 10 percent in some years. For a small economy such as the Philippines, a 10-percent increase in electricity consumption was very significant. Meanwhile, in the early Sixties, GPU decided to sell its interests in Meralco to a group of Filipino investors led by Eugenio Lopez Sr. and other prominent people known for their interest in the sugar industry. On January 5, 1962, the Meralco Securities Corporation (organized by Lopez Sr. and Alfredo Montelibano) purchased Meralco from its American owners with the aid of an US$ 11.9-million loan from the Philippine National Bank. The utility firm came under an all-Filipino board led by Lopez Sr., Salvador Araneta, and Luz Magsaysay, wife of President Magsaysay.

Eugenio's younger brother Fernando was the Senate president pro tempore at the time of the takeover.[82]

The National Power Corporation was one of the many government-owned and -controlled corporations (GOCCs) created from the American era to independence in order to spearhead the country's drive toward industrialization. But by the late Sixties, some quarters began questioning the existence of GOCCs, not only because many had built a reputation for bad management and insolvency, but also because they were perceived to be crowding out the private sector. Although various administrations since the Commonwealth justified the existence of government-owned corporations on grounds of public interest and national defense, inefficiency, corruption, and corporate losses seemed to outweigh whatever contributions these GOCCs made to national development.[83]

Yet, even the staunchest critics of GOCCs conceded that the National Power Corporation was an exemption. Since its rehabilitation in 1945, NPC's assets had grown from Php 11 million to over Php 700 million by 1968, with a net worth of over Php 380 million. Although limited to generating electric power and selling this in bulk to Meralco, provincial utilities (e.g., Davao Light and Power, Cagayan Electric) and electric cooperatives, Napocor's steady buildup of generating capacity enabled it to eventually supply 40 percent of the country's total power requirements (with Meralco providing almost 60 percent). And despite the fact that all of its projects had to be financed from foreign and local borrowings, the corporation's asset-to-liabilities ratio was four-to-one, much better than the two-to-one ratio generally accepted at the time.[84]

More important, NPC was fulfilling an important developmental role. Power generation was (and remains) an expensive endeavor that few private companies could afford. The construction of power plants often

---

82. Alfred W. McCoy, "Rent-Seeking Families and the Philippine State: A History of the Lopez Family," in Alfred W. McCoy (Ed.), *An Anarchy of Families: State and Family in the Philippines* (Quezon City: Ateneo de Manila University Press, 1994), 504-505.

83. "The government corporations: The P5 billion business (First of a series)," *The Sunday Times Magazine*, February 23, 1969.

84. "Let's wish the NPC more power (Second of a series), *The Sunday Times Magazine*, March 2, 1969.

necessitated foreign loans that required some kind of government guarantee one way or another. At the same time, to encourage agro-industrial development, the government had to promote the electrification of rural areas. But despite offering concessional loans, the government could not convince private utility firms to sink capital in areas that had insignificant markets. Naturally, National Power had to fill the gap, but it was not easy. Even back then, the company already experienced cash-flow problems. Since internal cash generation was inadequate to finance the peso requirements of foreign loans, NPC had to rely on local borrowings and government assistance to cover the shortfall. But it was definitely making money from operations and was among the few solvent government corporations.[85]

Toward the end of the Sixties, Napocor's experience had yielded sufficient lessons about the vulnerability of hydroelectric power to climate changes, particularly droughts. Thus, besides building oil-fired thermal plants, National Power began to explore nonhydro and non-oil-based sources. In August 1970, President Marcos directed the company to undertake commercial exploration and exploitation of geothermal energy. This eventually led to the meeting in California in 1970 that I set up between Executive Secretary and NPC Director Alejandro Melchor and Union Oil Company of California's Fred Hartley.

Meanwhile, Meralco continued to grow under Filipino management, building more oil-fired power plants to service the fast-growing Luzon market. In 1967, Meralco even explored the possibility of setting up a nuclear plant to service Manila's needs and commissioned a feasibility study for this purpose. Meralco's expansion eventually had serious financial consequences, as the company's long-term, dollar-denominated debts increased. The situation became worrisome for Meralco in 1970, when the Marcos government floated the exchange rate, resulting in increased cost of production of electricity and higher interest payments for foreign debts. This undermined the company's viability and it had to petition a 37-percent power rate increase in 1971. In May 1972, Meralco

---

85. *Ibid.*

requested for another 36.5 percent rate increase, but the Public Service Commission (the precursor of the Energy Regulatory Commission) approved only a 20.9 percent rate increase. This exacerbated Meralco's financial problems.[86]

In March 1971, the government formulated a twenty-year power development program (1971-1990) that envisioned the construction of thermal, nuclear, and pump-storage hydro plants for Luzon; diesel and hydro plants for the Visayas; and hydroelectric and thermal plants for Mindanao. The program also included electrification projects involving NPC, the National Electrification Administration, and the electric cooperatives.[87] On November 7, 1972, the President issued PD 40 ("Establishing Basic Policies for the Electric Power Industry"), which gave NPC a monopoly in power generation and transmission, and mandated the company to set up transmission line grids and associated generation facilities in Luzon, and major islands in the Visayas and Mindanao.

The government had to resort to PD 40 because the country reached a point in which creditors did not entertain loan applications unless these were accompanied by sovereign guarantees, given the magnitude of the loans. Now, if the borrower was a private company like Meralco, why would the government put up a guarantee? Given that the power sector was the single biggest borrower, the government decided to take control since it would have to guarantee the sector's loans anyway.

Under the new setup, private utilities, cooperatives, local governments, and other state-authorized entities could still participate in the electricity sector, but only in the distribution of Napocor-generated electricity.[88] This had far-reaching consequences for Meralco. Whereas it used to own power plants that provided a substantial portion of its total power sales, Meralco was now limited to the purchase of NPC's electricity for distribution to franchise areas in Greater Metro Manila and nearby provinces. From a business standpoint, PD 40 limited Meralco's potential for further growth

---

86. Ignacio, 20-21.
87. Ibid.
88. Presidential Decree 40, "Establishing Basic Policies for the Electric Power Industry," November 7, 1972.

since the company now had to rely solely on distribution of electricity for its profits.

When I became Napocor's chairman ex officio in 1977, I knew that it was a very different corporation from PNOC. The latter had minimal capitalization, but managed to generate enough profits and had other credit sources to finance its projects. On the other hand, NPC was heavily dependent on loans and government capitalization, and had other serious problems. In the late Thirties, the company had competent people; in fact, its engineers later built a reputation as among the best designers of hydroelectric systems in Asia. Over the years, however, the organization got fat and, especially after the war, the National Power Corporation began to evolve into a typical government bureaucracy.

Corporate directors assumed roles and functions that were beyond usual practices. The NPC employees had imbibed the usual government employee mentality, and lacked the professionalism and discipline of their PNOC counterparts, most of whom were trained under Esso's corporate setup. I also felt that the NPC management was inadequate for the tasks that the company laid for itself.

Thus the first item on my agenda as NPC chairman was to reorganize the company. Through PD 1360, which amended the organization's charter,[89] we declared all positions vacant, upgraded the position of general manager to president and CEO to better reflect executive responsibilities, and specified the duties and responsibilities of the board of directors. More important, I brought in key people from PNOC. I initiated a memorandum of agreement (MOA) between PNOC and NPC wherein several PNOC executives would be seconded to the Napocor to provide managerial, technical and financial expertise necessary for improving Napocor's operations. In return, PNOC received a monthly payment of Php 120,000 in management fees. Since PNOC had a better salary structure, the MOA enabled me to transfer PNOC executives to Napocor without them having to take a salary cut.

---

89. Presidential Decree 1360, "Further Amending Certain Sections of Republic Act Numbered Sixty-Three Hundred Ninety-Five Entitled 'An Act Revising the Charter of the National Power Corporation'," as amended by Presidential Decree 380, 395, 758 and 938," April 25, 1978.

Chapter 4: Privatizing Power Generation—The Tale of Napocor • 141

Through this MOA, which took effect on December 1, 1977, and was to run indefinitely, I was able to detail six PNOC executives and managers to assume key positions in NPC. Among these were Jose U. Jovellanos and Antonio L. Carpio, who were appointed senior vice president for engineering and vice president for finance, respectively (in 1980, Federico E. Puno, our senior finance analyst at PNOC, replaced Carpio). In reorganizing management, I emphasized the financial side. After all, Napocor was the biggest borrower among government corporations, and it was important for the organization to have competent financial managers. That's why I deployed some of the best finance people from PNOC. However, although I did not anticipate it at the time, this arrangement later created a class distinction of sorts within Napocor's management staff, since the PNOC men enjoyed higher salaries compared to the executives directly hired by the former, including the president.

I was fully aware of the irregularity in this situation. Actually, during President Marcos's time, the Commission on Audit had already questioned the MOA. The auditor even wanted to look at PNOC's salary structure, which was already beyond his authority because the company was not subject to government audit, and I told him so. I also explained that I had President Marcos's authorization to make that MOA. I suggested to the auditor to make representations with the president that we should nullify the MOA and open PNOC's books for government audit; he relented. If I did not have President Marcos's support, I can imagine the kind of compromises I would have had to make to pacify the auditor.

I grant that, from an auditor's standpoint, the MOA was "anomalous" but, without it, I would not have convinced the PNOC executives to transfer to NPC. Yet, because of their experience, these PNOC executives actually boosted the management capability of National Power. Sure, what I did was unconventional and perhaps by today's standards "illegal," and only martial law made it otherwise. But considering the state of Napocor today, I would prefer to resort to such necessary shortcuts compared to how subsequent administrations have run the company. We never lost money! When it was necessary to raise power rates, we did not hesitate to bite the bullet, unlike our successors who allowed political considerations to undermine NPC's viability.

At any rate, in April 1978, President Marcos appointed Gabriel "Gabby" Y. Itchon as Napocor president upon my recommendation. Gabby was then presidential assistant on financial and monetary affairs, and had been NPC director since September 1976. He had good credentials. Besides being a mechanical engineer, he had a master's degree in statistics from the University of the Philippines and a master's degree in economics from Yale University. He had worked for a long time at the Central Bank before his appointment as undersecretary for trade and industry in 1974. As my fellow director at NPC in 1976, I saw that he had good ideas for making the company more efficient and effective. We also worked well together, and so I thought he was the man needed to shake up National Power, so to speak.

Together with Gabby's appointment, President Marcos also increased Napocor's capitalization to Php 50 billion. In November 1978, the company purchased Meralco's Sucat, Gardner, and Snyder power plants for Php 1.1 billion, of which Php 400 million came from the Oil Industry Special Fund under LOI 735; the said amount was treated as government's additional equity contribution to NPC. Four more power plants and a 230-kilovolt transmission line were bought in January 1979; I myself authorized the release of Php 900 million for the purchase in my capacity as minister.

The addition of Meralco personnel increased our staff at a time when I wanted to trim down the work force, which numbered more than 7,000. Since Napocor employees were covered by civil service rules, we could not retrench excess personnel as easily as we did in PNOC in the early years. So one of the things Gabby and I worked out was a fairly drawn-out redundancy program that would reduce the work force to about 5,000. I actually wanted it cut to 3,000; after all, PNOC had the same number and yet it was running many more projects compared to NPC. But Gabby persuaded me that 5,000 was necessary to meet the organization's commitments. Unfortunately, instead of a reduction, the work force even grew to 10,144 by the end of 1978, and peaked to 12,062 at the end of 1981, before gradually tapering off to 10,564 at the end of 1985.[90] The significant increase in personnel was due to the fact that National Power had to work on about twenty different projects, including the nuclear power plant.

The company was also having difficulty with several of its projects. When I got appointed as energy minister, I did not expect that I would be experiencing serious political challenges in the course of doing my job, especially as National Power's chairman ex officio. For the first time, perhaps since I joined the government in 1973, I realized that implementing a development project could cost lives. Besides the Bataan nuclear power plant, NPC was implementing another politically controversial project, this time in Kalinga Apayao and Mountain Province—the Chico River Basin Development Project.

The Chico River is the longest and most elaborate river system in the Cordillera mountain ranges in Northern Luzon. In 1974, Napocor proposed the construction of four dams, with a total capacity of 1,010 MW, to be built with funds from the World Bank. The Chico River hydroelectric project would have completed the list of hydro projects for Northern Luzon. Site investigation had actually begun in the Sixties or even the late Fifties, but it was only in the Seventies that the project gained momentum.

When I became minister, Napocor was already encountering serious problems in the preconstruction phase of Chico IV, which was the largest of the four dams planned under the project. The preconstruction phase alone already had seven officers-in-charge by 1978, and resistance to the project became more organized after the tribal leaders of affected villages had a *bodong* (peace pact). NPC personnel could not even set up camp in the project site without bringing in escorts from the Philippine Constabulary (PC) because of the natives' constant interference and harassment.

---

90. National Power Corporation, 1985 Annual Report (Quezon City: National Power Corporation).

From what I know of the project's history and from what NPC people told me, the problem in the Cordillera was exacerbated by the policies of Manda Elizalde, whose task it was as Presidential Assistant for National Minorities (PANAMIN) to look into the protection of affected cultural communities' welfare, and to help ensure proper compensation and relocation. But it seems that the Kalinga villagers viewed Elizalde's policies as geared primarily toward promoting his private business interest, particularly in obtaining mining claims in the area.

I met Elizalde when I was still with Dole. He built his reputation in PANAMIN especially when he claimed to have discovered a "lost" tribal group in Mindanao called the Tasaday (there were many questions as to whether the Tasaday was indeed a "lost" tribe). He acted like a father and protector to the national minorities. The first time I saw him among these people, I could not help but marvel at the sight of these groups lining up to kiss his hand. But apparently, the Kalinga people led by Macli-ing Dulag suspected that mining claims, not their welfare, was what Elizalde was after. The villagers grew more determined to block the project, and the tension escalated when Elizalde responded by bringing in the PC and his own armed men.

I think it was his handling of the situation that really galvanized village opposition and raised the stakes in the Cordillera. The New People's Army (NPA) entered the picture, and National Power's survey team and their military escorts increasingly experienced ambush and raids on their camp. By the time I came on board at NPC, things had gotten worse and there were steady reports of company personnel getting killed.

When I first visited the project site, my PC escorts warned me not to take any food or drink from the local store, lest I "become sick." They also warned me that the villagers were ready to lose their lives in resisting the project. When I met with Macli-ing sometime in 1978, he told me about how the land had belonged to his people for generations and that the government had no right to take it away from them. But what I cannot forget was when he planted a certain flower and told me, "As my father used to say, if that flower grows, then it is my duty to kill you." Several months later, the flower blossomed!

By that time, the Chico river project site had turned into deadly terrain and it became increasingly untenable to pursue the project. Of course, it did not help that the anti-Marcos opposition used the project in their campaign against the president. The Manila-based opposition also pressured the World Bank to withdraw its financial support for the project. Toward the end of 1978, I recommended to the president the suspension of the Chico river project; it was simply not worth losing lives for, and I was not going to gamble with people's lives just to have the project done. The only question President Marcos had was whether we had any alternative to the planned Chico river dam, and when I assured him that we had plenty of options, he approved my recommendation. We proceeded to withdraw the NPC personnel assigned to the area. In subsequent months, I found out that military operations in the area persisted after we had pulled out, leading to the assassination of Macli-ing on April 24, 1980, and the intensification of the armed struggle.

Personally, I felt sorry that the project could not push through. The Chico river is the only known natural drain for rainwater, which is why you need not build a massive dam just to set up your hydroelectric plant. This means you can generate a cheap source of power because your first cost is much less. Those advocating the rights of national minorities claimed to protect the interest and heritage of the minorities there. But I think that is a myopic perspective, because the project was meant to serve the national interest, which was to provide a cheap source of electricity. And that national interest not only affected the majority of Filipinos, but also included the Cordillera villages; after all, they are fellow Filipinos. Unfortunately, the anti-Marcos groups saw the project as symbolic of the conflict between national and minority interests, and of President Marcos's "developmental aggression" and violation of the rights of indigenous peoples. I am not denigrating the Cordillera people, but I think we missed a great opportunity for the nation.

Generally speaking, I think we have to realize that no matter what you do, in promoting structural development such as dams or other major infrastructure, there are social consequences. I remember when I visited Lanao del Norte to inspect the construction of Agus-7, the Air Force helicopter I rode in could not land because we were being shot at from somewhere in the forest! Since damming up the river necessitated the inundation of large tracts of land, it was inevitable that some villages would be affected. This, in turn, sparked community resistance. Gabby Itchon once told me about a local Muslim resident who confronted him at the Agus-7 project site and said, pointing to the land, "My father is buried there. You submerge that, and I'm going to kill you!" Pursuing the project essentially meant seeking military escorts to secure personnel and equipment. Despite this, the Agus-7 project claimed at least fifty lives among NPC personnel.

I think it is also important to remember that Napocor's projects were located in places where there were active insurgent movements. Even in the late Forties and early Fifties, company personnel in Central Luzon had traumatic encounters with the Huks. With the expansion of the NPA and other armed groups opposed to the Marcos government, NPC could not avoid getting caught in the crossfire. For instance, when the company was building a power plant over the natural gas well that PNOC drilled in Cagayan Valley, the project's completion was delayed because the NPA were ambushing National Power's engineers! This vulnerability to armed groups forced the organization to rely on military support. But, at least during my time, I do not recall having had to deploy the military against unarmed resistance by local people. My attitude had always been that I would not insist on a project which is not welcomed by the community.

I remember an instance during the construction of the Palinpinon geothermal plant in Negros Oriental. On my way back to Manila, students from Silliman University accosted me at the Dumaguete airport for not performing an environmental impact study at Palinpinon, which I grant. It was true, but only because an environmental impact study would take quite a while and I did not have time to spare. We did many of these projects without considering an environmental impact study, which is not correct according to theory. But what do we do, wait? So I told the students, "If you think we should have an environmental impact study, we'll pull out tomorrow, there's no problem. We'll go somewhere else." They did not respond.

Despite these various challenges, the National Power Corporation was viable, and my intention was to strengthen it further and make it a stand-alone company that relied less on government equity contributions to shore up its capital position. To this end, we instituted the following basic financial policies:

- First, Napocor should make money from selling power.
- Second, it should make enough to fund its peso capital expenditure rather than rely on borrowings.
- Third, it should make a minimum of 8-percent return on rate base (RORB).

The last policy was actually in compliance with the condition of the World Bank and the Asian Development Bank, Napocor's principal creditors, that the organization meet the peso requirements of its projects from internally generated funds. The 8-percent RORB was designed to help NPC make the necessary margins.

The RORB had already been part of the company's financial policies even before I became chairman; the initial RORB was 6 percent. NPC's foreign creditors imposed the RORB to make sure they got paid and also made a handsome return. The 8-percent rate is actually an arbitrary figure, decided upon at a time when the world was still experiencing low inflation. Immediately after the war, from the Fifties to the early Seventies, inflation was about 2-3 percent. It was only in the Eighties that inflation shot up to as much as 20 percent in the United States, until Reaganomics came into play and they were able to bring down inflation to 7-8 percent. Frankly, imposing an 8-percent RORB on a country like the Philippines can only lead to greater hardship for our people. *Hindi na makakabangon ang Pilipino* (The Filipino will not be able to recover). Unfortunately, unless we imposed it, Napocor would be unable to borrow money from abroad. Where else would the organization get the funding? The 8-percent RORB was one of those conditions from international lenders that we accepted with much regret, knowing fully well that it would raise our electricity prices. There was nothing we could do.

In relation to the above policies, the Napocor board was authorized to approve the corporate budget, and to set the rates that would allow the company to make an 8 percent RORB; if NPC wanted to have a higher RORB, it had to go to President Marcos for approval.

We also formulated a ten-year Power Expansion Program (1981-1990) which emphasized the following basic thrusts: complete the backbone transmission systems in the country's major islands, meet the demand for electricity in these islands, and reduce the share of oil-fired plants from 63 percent of total output in 1980 to 30 percent in 1985. At the same time, the program's financing plan was designed to instill some level of financial discipline in NPC. First, the foreign-exchange costs of the power projects, including interest payments during construction, must be financed by foreign loans. This meant that Napocor could not begin project construction until after it had raised adequate financing of foreign-exchange costs through the international financial market. Second, the financing plan provided that government's equity contribution for the peso cost of power projects would gradually decline from Php 1.9 billion in 1980 to Php 1.1 billion in 1985, and would cease altogether once the projects had been completed.

Since the Power Expansion Program called for the accelerated development of non-oil energy sources, we projected that by the first half of the program (1981-1985), we would have substantially diversified our energy source mix. Our main alternative sources were hydro and geothermal, whose proven capacities were equivalent, respectively, to 18.2 million barrels of oil and 12.3 million barrels of oil on a yearly basis. From 1981 to 1985, the projected national energy source mix comprised the following:[91]

---

91. Jose U. Jovellanos, "Hydroelectric Power Development in the 1980's," *The Philippine Geographic Journal*, vol. 25, Jan.-Mar. 1981, 20.

Table 4.1
Projected National Energy Source Mix, 1981-1985

Source	In Percent		
	1980	1981	1985
Electric Power	37.99	34.08	39.64
Hydro	7.21	7.42	12.78
Geothermal	4.19	5.46	12.22
Coal	0.46	1.18	6.27
Oil/Diesel	21.13	19.93	5.25
Nuclear	—	—	2.10
Non-Conventional	—	0.09	1.02
Non-Power	67.01	65.92	60.36
Oil	66.39	64.55	49.70
Coal	0.59	1.20	7.14
Nonconventional	0.05	0.17	3.52

Under this projected national energy mix, we targeted the following electric-generation capacities:[92]

Table 4.2
Targeted Electric-Generation Capacities

	1980		1981		1985	
	MW	%	MW	%	MW	%
Hydro	933.0	20.8	959.0	20.4	2831.0	29.8
Diesel/Oil	3078.1	68.7	3229.0	68.0	3391.0	35.8
Coal	28.0	0.6	90.0	1.9	760.0	8.0
Geothermal	443.0	9.9	446.0	9.4	1726.0	18.2
Nuclear	—	—	—	—	620.0	6.6
Nonconventional	—	—	12.0	0.3	154.0	1.6
Total	4482.1	100.0	4736.0	100.0	9482.0	100.0

---

92. Ibid.

I am proud to say that from 1978 to 1985, National Power did well. Following its mandate under the Power Expansion Program, NPC completed the backbone transmission systems and supplied electricity to all fifty-six provinces and fifty-two cities in the country's seven major islands (Luzon, Mindanao, Panay, Negros, Cebu, Bohol, and Leyte) by 1985. Twenty new power plants that had been in various stages of construction were completed, save for Agus-3 in Lanao del Norte. From 1981 to 1985, Napocor had new plants with a combined capacity of more than 2,000 megawatts, and almost 90 percent of these came from non-oil sources. In keeping with the president's mandate to reduce the country's dependence on oil, the company reduced the share of oil-based power plants from 64 percent of total power output in 1980 to 36 percent at the end of 1985. We could have reached our original target of 30 percent had the completion of the nuclear plant and two hydro systems in Mindanao not been delayed.[93]

Although the government did not approve the Power Expansion Program until 1982, which delayed the implementation of rate increases, Napocor was still selling power at a price higher than its production cost. In fact, as the company's 1985 annual report indicated, "Between 1977 and 1985, sales revenue had gone up from Php 0.4 billion to Php 18.0 billion" while total assets had grown to Php 107.2 billion, almost ten times more than what it was in 1977. And after 1978, when President Marcos authorized the company to raise its capitalization to Php 50 billion to be financed through government's equity contribution (part of which NPC used to buy Meralco's power plants), there were no more increases in capitalization from the national government. There was never any need to request the Batasang Pambansa (National Assembly) for additional government equity contribution since Napocor managed to live within its own corporate cash flow.

Of course, National Power also benefited from the support of PNOC, the most profitable government corporation at the time. PNOC had been extending generous short-term oil credits to NPC, and the organization had to rely on these credits even more during the financial crisis of 1983-

---

93. National Power Corporation (1985), "Report of the President," 1-4.

84. By the end of 1985, Napocor had unpaid fuel oil purchases from PNOC worth more than Php 1 billion.

The severe economic crisis that followed the government's declaration of a moratorium on foreign debt-service payments in October 1983 affected Napocor since its creditors cancelled or suspended foreign loan contracts from October 1983 until the latter part of 1985, forcing NPC to defer construction on new projects. Other stabilization measures, the rapid rise in inflation, and the continuing slide of the peso against the dollar added to the organization's financial problems as its costs rose rapidly. In 1983, NPC's cost of power service was Php 0.5790 per KWh; by 1984 it increased by nearly 40 percent to Php 0.8754, and settled at Php 1.0984 at the end of 1985. Yet, National Power performed creditably. In 1984, the company had a net income of Php 1,068,689,471; this increased slightly to Php 1,096,292,079 in 1985.[94]

At this point, let me stress that we made it a matter of policy to actively invite various sectors like the media and business groups to scrutinize NPC's books. There was never any attempt to smooth over problems or tamper with accounts, especially during the economic crisis. I told Gabby Itchon to provide a detailed picture of Napocor's problems and financial standing in the annual reports[95] because this was the only way we could avoid suspicion. This policy of transparency in reporting and accessibility of company records, which I had imposed on PNOC and its subsidiary companies at the onset, and later at the ministry, was my safeguard against accusations of corruption.

I also want to emphasize that all purchases for fuel, equipment, and power plants followed the stringent and transparent international bidding procedures of multilateral and bilateral financial institutions that were far stricter than government bidding procedures and required the concurrence of the funding agency in the award of the contract.

I can very well say that as chairman, I succeeded in making Napocor a stand-alone company, one that could generate income from operations,

---

94. Ibid.
95. As a matter of policy, the annual reports of the Ministry of Energy and all attached agencies and corporations, including PNOC and Napocor, were published and made available to interested parties.

without having to depend on government subsidies. Although the nature of its mandate made it difficult for National Power to match the profitability of PNOC, the organization certainly showed that, properly administered, it could maximize its income potential.

When the Aquino government came to power in February 1986, I was succeeded as NPC chairman by Edgardo Espiritu, who had made his name in banking, while Gabby Itchon was replaced as president by Conrado "CD" del Rosario. CD actually began his career in Napocor in 1944 as an electrician and rose up the ranks to become general manager from 1975 to 1978. He was NPC's signatory to the contract with Westinghouse for the nuclear power plant.

In a span of less than six months, the Aquino government made a series of important decisions that had far-reaching implications for National Power. The first significant decision was to mothball the nuclear plant, which meant that the organization continued to amortize its loans without generating a single centavo in income. This situation went on until June 1986, when the government formally ordered Napocor to stop making any payments related to contract obligations or foreign loans.

Also in February, the government ordered National Power to give "discounts," a form of rate reduction, to customers using the Negros and Mindanao grids and the Cebu sub-grid, purportedly because of the decrease in oil prices during the year. The government also delayed the incorporation of the Currency Exchange Rate Adjustment (CERA) in Napocor's billing charges until November 1986; we introduced the CERA in 1985 to enable the company to cope with the unstable dollar-peso exchange rate, and this was supposed to take effect beginning January 1986. As National Power itself reported, these two decisions alone on rate reduction and the CERA caused an 8.2-percent reduction in NPC's net revenue from sales to Php 16.514 billion as against Php 17.991 billion in 1985, *despite* a 2.9-percent rise in energy sales for the year. The RORB also went down to 7.4 percent compared to 8.2 percent in 1985.[96]

---

[96] National Power Corporation, 1986 Annual Report (Quezon City: National Power Corporation), 2-3. Unless indicated otherwise, references to NPC's situation in 1986 came from the said annual report.

More important, Napocor's financial situation in 1986 began to show signs of things to come. Although the company reported an 18.8-percent increase in gross internal cash generation, from Php 6.795 billion in 1985 to Php 8.074 billion, this was offset by debt-servicing commitments amounting to Php 8.148 billion. As a result, NPC incurred a cash deficit of Php 6.376 billion in its operating and capital expense accounts. This deficit was financed, in turn, by foreign borrowings (Php 2.335 billion), advances from the national government (Php 1.257 billion), deferment of debt-service payments under a loan restructuring program (Php 2.105 billion), and other domestic sources (Php 679 million). From 1985 to 1986, National Power suffered a 104.4-percent decrease in net internal cash generation from Php 1.699 billion to negative Php 74 million. All this combined to reduce NPC's net income to Php 715 million, a 34.8-percent decrease from the 1985 level of Php 1.096 billion.

What happened? How could a company that was operating respectably under very adverse conditions in 1985 suffer such drastic reversals in just one year, at a time when the economic crisis had eased somewhat? The National Power Corporation's 1986 report essentially blamed debt servicing. I will grant that it was an important factor, but I believe the other decisions concerning the rates and the CERA were also important. It is also interesting to note that NPC's personnel increased from 10,564 in 1985 to 10,821 in 1986, purportedly for the repair of existing plants and transmission lines, besides the construction of new ones. In the next few years, personnel enrollment reached a high of 14,000 as the Napocor management accommodated thousands of political appointees despite the fact that the organization was not doing much.

With the creation of a new constitution and the restoration of Congress in 1987, the new regime reversed some of the martial law-era regulations that had benefited the operations of National Power and even PNOC. Republic Act 7638 ("Department of Energy Act of 1992"), which reestablished the Department of Energy, required the NPC, PNOC, and the National Electrification Administration to submit their respective budgets for congressional approval. This law also repealed the provision in the PNOC Charter making the energy secretary the concurrent CEO of the company.[97] Likewise, under the Aquino administration, it was stipulated that decisions on Napocor rates must be approved by a body higher than the board of directors—initially by the President of the Philippines and later by the Energy Regulatory Commission.

These changes opened the door to political interference in NPC's operations. In the post-Marcos years, Mrs. Aquino actually began the practice of Philippine presidents holding back rate increases by Napocor or even reducing existing rates for political purposes. From 1988 to 1991 alone, after NPC was ordered to peg power rates in rural areas to Php 2.50 per KWh when its cost was about Php 4.50 per KWh, the company shouldered Php 81 million in generation subsidy. The Ramos administration continued the practice of politicizing Napocor's rate setting to a point that the company absorbed Php 1.36 billion in generation subsidies from June 16, 1988 to December 31, 1995.

In the meantime, the government did not plan for any alternative power source that would replace the 620 megawatts that the nuclear plant could have provided. People kept pointing to the project's US$ 2-billion cost, yet because of Mrs. Aquino's decision to mothball the plant without planning for the replacement of its 620-megawatt output, we suffered daily power outages of ten to twelve hours for almost four years! How many billions of dollars did that cost the economy?

My concern then was that the Aquino government totally discredited the energy sector. The Ministry of Energy had been dismantled and nobody took responsibility for energy security. In my first public appearance in 1987, on the late Louie Beltran's television show, I made a statement that unless the government moved fast to put up new power plants, the country would run out of electricity in two years' time. True enough, in 1989 we started having daily power outages that lasted until 1993, which forced the government to undertake the expensive but ultimately shortsighted stop-gap measure of contracting independent power producers (IPPs). These were companies that brought in power barges or built their own power plants, and sold electricity to the government at a guaranteed price,

---

97. Republic Act 7638, "Department of Energy Act of 1992," December 9, 1992.

regardless of whether it is used or not. It is a virtual guarantee for profit at the expense of government and the consumers. No wonder that this matter of IPPs is vulnerable to suspicions of corruption.

The mothballing of the nuclear plant coupled with inadequate maintenance of aging plants and the declining capacity of hydroelectric plants made the country's energy picture worrisome. But nobody in government paid heed, and so by 1989, the country experienced a power crisis marked by eight to twelve hours of power outages. These daily brown-outs—which lasted until 1993—were estimated by the World Bank to have cost the country unemployment and economic losses equivalent to 1.5 percent of GDP every year; the business sector put the losses at about US$ 1.3-billion.[98]

But the government did not do anything to address the situation until June 1991, when Mrs. Aquino appointed my former colleague Pablo "Pabling" V. Malixi as NPC president. Upon his appointment, Pabling presided over a "fast-track program" that assembled nineteen power plant projects through investment schemes like build-operate-transfer (BOT). Some of these power plants were put onstream during his term (which lasted until April 1993), while the others were inaugurated a few years afterward. I know that Pabling tried his best to address the power crisis, considering the odds he faced. But I think he was overwhelmed by the problem since his "fast-track program" resorted to contracting independent power producers (IPPs), which we now regret. I would have expected Pabling to temper his plans with more ingenuity, rather than simply taking on an approach that was inappropriate and shortsighted to begin with.

Mrs. Aquino's successor, President Fidel V. Ramos, responded to the crisis by using the emergency powers that Congress granted him in April 1993 to approve contracts with IPPs for the construction of power plants whose output Napocor would buy. Most of these contracts had a "take or pay" provision wherein National Power was required to pay for most of the power generated by these IPPs, whether or not the organization needed

---

98. Natsuko Toba, "Welfare Impacts of Electricity Generation Sector in the Philippines," Asian Development Bank ERD Working Paper Series No. 44, June 2003, p. 2.

this. To ensure a market for the IPP-generated power, NPC entered into a ten-year contract with Meralco in 1995 for the latter to buy 3,600 megawatts of power from Napocor until November 2004. But Meralco was also building its own IPPs, from which it eventually bought its electricity supply. The situation had become like a free-for-all. Meralco stopped buying power from NPC in 2001, despite the fact that the purchase contract had yet to lapse, leaving Napocor with substantial reserves that had been contracted precisely for Meralco's purchase requirements.

As a consequence of the economic slowdown following the 1997 Asian financial crisis, there was a reduction in the country's power consumption. This situation, in turn, contributed to a glut in power supply toward the end of President Ramos's term in 1998. Unfortunately, Napocor had about forty IPP contracts to honor despite the glut. Consumers shouldered the added costs from these contracts by paying for the notorious Purchase Power Adjustment (PPA) built into their electric bill. Over the next several years, NPC deteriorated to the point that, by the end of 2004, the organization was reported to have a financing deficit of Php 73.4 billion, which the Arroyo government hoped to trim by privatizing the remaining Napocor assets like transmission lines and the power plants.[99]

The once-proud government corporation of the Fifties has now become a worthless organization. The present government had hoped to get some money when they bid out the company's assets, but so far, very few investors are interested. But more than generating funds by privatizing NPC, the government seems to be forgetting that the principal objective should be how to obtain lower electricity rates. Leaving power generation to private investors will not lead to cheaper electricity because no private investor will agree to sell power unless there is a mark-up, which will surely be considerably higher than Napocor's. The latter could afford a lower mark-up because it pays lower interest rates on loans, which are government-guaranteed.

If I were to trace the origins of Napocor's downfall, I will have to go back to February 1986. In terms of policy, two serious faults that Mrs.

---

99. "Asset sale to cut financing deficit of Napocor to P4.165B," *The Philippine Star*, May 13, 2005.

Aquino made in relation to the energy sector were the mothballing of the nuclear power plant without planning for replacement plants, and the deferment of NPC's power rate increases due to political considerations.

But there was another decision made by Mrs. Aquino that affected National Power's primary role as the country's sole body authorized to generate electricity, and this was EO 215, which privatized the generation of electricity.[100] Among other things, this law enabled NPC's historical "competitor" and "client," Meralco, to go back to the business of power generation, on top of distributing electricity.

Before martial law, Manila's electricity services were essentially handled by the Meralco Group, which originally had two components: a distribution system that included the electric posts, transformers, electric meters, etc., and a power-generation system wherein it operated its own power plants and supplied most of Manila's electricity, with the National Power Corporation supplementing the remaining requirement. The firm's franchise operations covered not only Metro Manila and most of Central Luzon, but also parts of Southern Luzon.

When President Marcos issued PD 40, power generation was centralized under Napocor, and private utility companies like Meralco were limited to distributing NPC-generated power to their franchise areas. This changed when Mrs. Aquino issued EO 215, which allowed various types of organizations to generate their own electricity, in effect breaking up Napocor's monopoly.

But Mrs. Aquino made another ill-considered decision when it "returned" Meralco's electricity distribution system to the Lopez family, without reflecting upon what was best for public interest and without giving a thought to alternative institutional arrangements that could best promote the welfare of Meralco's consumers.

My main concern is that Meralco already enjoys a monopoly over the premier power distribution market; why allow its related companies like First Philippine Holdings Corporation to go into power generation?

---

100. Executive Order 215, "Amending Presidential Decree 40 and Allowing the Private Sector to Generate Electricity," July 10, 1987.

I strongly believe that, as a matter of policy, power generation and power distribution are distinct and separate activities, and no single company or business group should simultaneously engage in both. Since power generation and distribution have a bearing on public interest, it would be ideal for these two endeavors to be subject to government regulation. Unfortunately, from the time of Mrs. Aquino until the present, the government has steadily moved toward deregulating and privatizing the power sector, culminating in the passage of Republic Act 9136, or the "Electric Power Industry Reform Act of 2001" (EPIRA).

EPIRA opened the door to full privatization of electricity generation and transmission, effectively sounding the death knell of National Power Corporation by mandating the sale of its generation assets, limiting the organization to missionary electrification functions, and creating the National Transmission Company that would assume Napocor's transmission functions.[101] More important, by allowing private companies to produce power without having to obtain any franchise, the government is effectively treating power generation as an ordinary business rather than as a public utility.

When a company that is producing power (an open, competitive, and deregulated business) is also involved in electricity distribution (characterized as a public utility and therefore requiring a franchise), the potential for conflict of interest is enormous. The distributor will always choose to buy from his related generator at the highest price the market can bear, or allow his related generator the right to match the lowest offers from others (who may or may not offer the lowest price available in the market). Thus, the exercise becomes one of price matching and not of seeking the least cost. This defeats the purpose behind the competition policy in the power-generation sector. The generator and distributor should be separate and distinct in order to promote true and transparent competition. With due respect to our legislators, I am afraid that such a situation was not thoroughly studied when EPIRA was being crafted.

---

101. Republic Act 9136, "Electric Power Industry Reform Act of 2001," June 8, 2001.

Considering that the distribution of electricity is a public good—which is the rationale behind the constitutional requirement of a franchise—we must ask ourselves whether the present setup actually serves the common good, which is providing the least cost of electricity to consumers. We must remember that the cost of electric power, especially in urban areas, is a serious factor in determining our global competitiveness.

I firmly believe that one cannot be a producer of electricity and be its distributor as well. This is a formula for conflict of interest. I cite the case of the Lopez family because it is the most prominent player in the power industry, but there are others like Visayan Electric Company in Cebu or Davao Light and Power that are similarly situated.

But even assuming that, in principle, there is nothing wrong with companies generating and distributing power at the same time, the government should periodically review the performance of these companies insofar as distribution is concerned because, after all, they are doing so under a franchise. Again, let me cite the case of Meralco. Considering that its franchise areas are in urbanized sectors, it is ironic that Meralco probably has one of the oldest electricity distribution systems in Asia. In Southeast Asian cities like Jakarta or Bangkok, much less Singapore and Kuala Lumpur, you will not find "buntings" of transmission wires, rotting wooden electric posts, and transformers hanging from the posts that you find even in the most exclusive subdivisions in Manila. Meralco should really gear its system toward installing underground cables so that these are protected from lighting and other hazards borne by open-air exposure.

Meralco should invest in updating such an antiquated system. In fairness, when Lopez Sr. was managing Meralco, the company delivered on its obligations efficiently and according to high performance standards. But the company has since been on a steady decline. The government should more fully exercise its regulatory powers in compelling Meralco to upgrade its system so that the interests of its customers are better served. After all, that is what the concept of a "franchise" is all about—that the holder is fulfilling a function imbued with public interest, and the government is granting monopoly rights to the franchise holder under certain conditions to ensure the protection of public interest. The Energy Regulatory Commission should monitor whether Meralco is actually fulfilling this requirement, besides deciding on petitions for power rate increases. The government needs to strengthen the commission to make it capable of detecting and correcting flaws in the system.

A related issue concerns corporate governance. Monopoly companies are often confronted by questions of transparency and accountability, and utility firms are especially vulnerable because theirs is no ordinary business. Utilities distribution involves the public good and is therefore a sensitive issue. A franchise utility like Meralco should be open about its operations and act with transparency on matters that have a bearing on consumer interest. This brings me to my second point vis-à-vis Meralco.

My own perspective is that Meralco should be owned by consumers, if possible. I remember that President Marcos asked me to prepare a plan whereby the consumers would eventually own Meralco, and the ministry did submit such a plan. I have not heard anything about it since I left government. Utility distribution is too critical and too sensitive a business. Right now, power distribution is a monopoly. The National Power Corporation was broken up in the name of dismantling government's monopoly, and yet we gave monopoly rights to another entity. This is effectively what the Aquino government did.

But it was not just NPC that suffered from poor decision making under the Aquino administration. The Ministry of Energy as a whole was dismantled, paving the way for the breaking up of an entire energy complex, probably on account of the lobbying by representatives of foreign oil companies. More important, Mrs. Aquino's so-called economic reform policies laid the groundwork for the eventual privatization of the country's most profitable government-owned corporation—Petron.

CHAPTER FIVE
ASSESSING the ROLE of FOREIGN OIL COMPANIES

W*hen the government bought out Esso (Philippines), the country obtained an excellent opportunity to gradually build its independence from the foreign oil companies. Through PNOC and Petron, the government developed a good knowledge of the oil industry and a strong capacity for competing with foreign oil companies in the retail market. More important, the government's entry broke the oil majors' monopoly and put these companies in a situation where they had to be more careful or judicious in their sourcing and pricing policies, which required a little more transparency on their part. Unfortunately, the Ramos administration sold 40 percent of Petron to a foreign company, Aramco. It was another ill-considered decision that would have far-reaching consequences for the country.*

5

GZV with Sheik Zaki Yamani, minister of oil for Saudi Arabia, the most influential member of OPEC. (June 1975)

(left) GZV with General Ibnu Sutowo of Pertamina during the signing of the US$ 150 million agreement to purchase oil from Pertamina. (January 22, 1975)

(bottom) GZV and T.E. Wallace, president of Esso Philippines Inc., signing document of sale. (December 23, 1973)

The hard work that all of us put in at the Philippine National Oil Company (PNOC) and the Ministry of Energy went down the drain on June 19, 1986, when Corazon Aquino issued Executive Order (EO) 20, abolishing the Ministry of Energy and placing all its attached offices, agencies, and corporations under the administrative supervision of the Office of the President. Although the EO seemed to imply that it was a temporary move "in the meantime that the Presidential Commission on Government Reorganization is still studying the overall structure of the government,"[102] it was the beginning of the dismantling of the country's energy infrastructure.

Why did Mrs. Aquino do it? In an interview with Sylvia Mayuga in 1991, her former executive secretary and now-Senator Joker Arroyo discussed the story behind the ministry's abolition. In the transcript of the interview, it appears that Mrs. Aquino abolished the ministry upon the advice of Cesar Buenaventura, who had claimed that the Ministry of Energy was "the most corrupt" among the Marcos-era agencies. This allegation was confirmed by Joker, who had witnessed how hard Buenaventura lobbied to have the Ministry abolished.

Now, Cesar Buenaventura was one of Mrs. Aquino's closest advisers, but he also happened to be the president of Pilipinas Shell at the time. I have no idea as to Buenaventura's basis for claiming that the ministry was the "most corrupt," but I also have no doubt that he had Shell's interest in mind when he recommended the ministry's abolition. I could sense that the foreign oil companies were never happy with PNOC, not only because Petron led the pricing structure in the oil market, but also because PNOC's energy development program, with its emphasis on tapping non-oil sources, threatened to erode the oil companies' position in the energy

---

102. Executive Order 20, "Placing the offices, agencies and corporations attached to the Ministry of Energy under the administrative supervision of the Office of the President," June 19, 1986.

market. Riding on the wave of anti-Marcos sentiment was a good way to eliminate a rival.

In my opinion, the abolition of the ministry showed Mrs. Aquino's inexperience in proper governance. Buenaventura may have been a close friend of hers, but how could she, in conscience, consult someone like him whose interest was to protect his employer, a foreign oil company operating in the Philippines? On the mere say-so of Buenaventura, Mrs. Aquino dismantled the whole energy complex that took twelve years to build and which, in government annals, was unique for the successes it achieved considering the constraints faced by the country. Incidentally, the Queen of England knighted Buenaventura thereafter. Did that have anything to do with the ministry's fate?

In the same interview with Sylvia Mayuga, Joker revealed that other advisers had already been eyeing Petron's privatization early on in Mrs. Aquino's term; they were lobbying for British Petroleum and for a Kuwaiti oil company. Other groups close to Mrs. Aquino's advisers were interested in PNOC's privatization because this would enable them to get their hands on Petron.

From a policy perspective, there was no reason to privatize PNOC/Petron even at the time. Why would a government in dire need of cash be willing to let go of a good source of income? PNOC was the biggest government corporation in terms of revenue. Much of it was due to Petron, which commanded about 40 percent of the local oil market and occupied the top spot in the industry. More important, as Joker himself acknowledged, PNOC's involvement in oil importation, refining, and marketing took away the foreign oil companies' advantage of being the only ones who knew how to play the game. It is not surprising, then, that Petron threatened the interest of multinational oil companies.

Joker's point about PNOC's impact on the oil companies is significant, if only because it affirms the fact that foreign oil companies have always invited suspicion that they act as a cartel and dictate the local price of oil regardless of international prices. This issue has hounded the local oil industry since the Ramos administration deregulated the sector, and has intensified in times of unabated oil price increases, such as what we are experiencing now. To my mind, however, there is one important question that we should confront—a question that has so far evaded a real answer: As a matter of policy, what should we expect from the foreign oil companies in the Philippines?

It was not until we operated Petron that I started to realize how critical an oil company is to a country. The oil companies in the Philippines have been with us for almost a century. In fact, the phrase "old China hand" emanated from the oil companies because some of those people assigned to the Philippines were "old China hands" who had been previously employed in the oil companies' pioneering ventures in China.

Judging from their history, I think we can safely say the oil companies' interest in the country is limited to business—any normal business venture: "I bring in the oil, I process it if there is a refinery, and then I will sell it to the country." I know that it is a very broad statement, since selling oil is not at all that simple. You have to have an infrastructure for selling—storage facilities, transport, and a marketing organization. If there is one thing that the foreign oil companies have done in a very efficient way, it is the *distribution of oil products*. No matter where you go—even to the smallest, remotest barrio or *sitio*—you will always find somebody having a supply of kerosene. The oil companies have developed a very efficient marketing and distribution network through the establishment of depots all over the country.

The question that you may ask of the oil companies is "What kind of profits are they entitled to?" And this is where the problem lies. Prior to Petron's entry into the market, all the oil companies were essentially foreign. Filoil had an effectively minority Filipino interest, as it was diluted by Gulf Oil of the US through its supply of crude oil. Although there was Filipino capital put into Filoil, it was not enough to affect the company's operation vis-à-vis the oil majors. So we can treat this industry as entirely international and foreign. Now, you will probably say that it is natural and reasonable for the oil companies to make profits, or else they will not stay in the country. But I submit that they have already made a lot of profits, and have been doing so for more than a century. *When is enough?*

It is true that they have put up investments in the Philippines. But if one were to make a serious study of how much we have really paid for in terms of capital replacement, it would be very hard to say that they have been reasonable to us. Probably the opposite might be true: We have been more reasonable to them than we have been to ourselves. Although this is a debate where the real answer may be very difficult to ascertain, I will dare argue that *none of these oil companies has ever lost money in the Philippines*. A reasonable rate of return? I think they have had more than a reasonable rate, especially in light of what we are experiencing today when crude oil prices hover around US$ 60 a barrel. But from my experience at PNOC, I think it is safe to say that whether there is a price increase or a price decrease, the oil companies always manage to make a handsome profit somehow.

In the early Seventies, when the government started regulating the oil industry, multinationals like Esso chose to pull out. The government ended up buying all of Esso's local assets, which included twenty-eight bulk plants, over 800 service stations in Luzon and Southern Philippines, and a 57-percent share of the Bataan Refining Corporation (Mobil Oil owned the remaining 43 percent).

Since the government already owned Filoil Marketing Corporation, the remaining foreign oil companies (Shell, Caltex, Mobil, Getty) became wary of government intentions when it took over Esso (Philippines). Remember that in the Sixties and early Seventies, British and American oil companies experienced problems in the Middle East when several governments nationalized their oil industry. With the consolidation of Filoil and Esso (Philippines) under Petrophil Corporation, the government effectively controlled about 35 percent of the oil market.

When I became head of PNOC, one of the first things that President Marcos warned me about was to be careful with the foreign oil companies. He knew the extent of political and economic influence that their mother companies wielded with their home governments and international lenders, and he did not want the creation of PNOC and Petrophil to be misinterpreted as preparatory steps toward nationalizing the oil industry. Although PNOC was tasked to secure the country's oil supply, the president's explicit instruction—which he kept repeating over the next several years—was that the government must limit itself to 40 percent of the local oil market. Although he did not say so, my sense of the president's position was that, between the fact that we were *the* government and that we controlled 40 percent of the market, our position was strong enough vis-à-vis the oil majors without having to monopolize the industry.

At that time too, when martial law was just over a year old, President Marcos was trying to build the new regime's legitimacy in the eyes of the international community. He wanted to make the Philippines a showcase of industrial development, and so it was important to send the right message to foreign business. Offending the foreign oil companies was certainly not going to help the country's image, much less encourage foreign investment.

The president's concern that Petrophil might encroach on the foreign oil companies' market share was well founded. In December 1974, or one year after the government took over, Petrophil was the country's third-largest corporation in terms of sales. Under the brand name "Petron," the company produced and marketed seven petroleum fuels; four types of motor oil and three branded gear oils, and transmission fluids for automotive lubrication; thirteen brands of industrial oils and three kinds of marine lubricants for industrial lubrication, two types of industrial and automotive greases; ten brands of process oils and six other special products, including rust preventives and protective coatings; four branded asphalt products; and various additives. These products were certified to meet the exacting quality standards specified by such international organizations as the American Petroleum Institute and the Society of Automotive Engineers.

Petrophil also had the most comprehensive bulk-plant coverage in the Philippines, with a presence in thirty-six strategically located points from Aparri in Luzon to Zamboanga and Davao in Mindanao. Support facilities included 290 contracted and company-owned tank trucks that could deliver on a monthly basis 862,000 barrels of petroleum products for use by service stations, LPG dealers, industrial consumers and government agencies. By 1975, Petrophil had the most extensive retail network in the country, with 1,173 service stations catering to the needs of the motoring public. We achieved this by combining and rationalizing Esso (Philippines) and Filoil's facilities. We also set up Petron Tires, Batteries and Accessories (Petron TBA) to make our gas stations more competitive because of the additional products and services (like the Car-Check Center) they provided.

To further support the marketing effort, quality-control laboratories at Pandacan and at the Bataan refinery assured the quality of Petrophil's products. These laboratories were equipped with sophisticated petroleum-testing equipment and were operated by a highly competent staff of technicians. A technical service department provided training and expert advice on petroleum application and usage to government and industrial accounts.

Let me digress at this point. One of the first things we did upon taking over Esso (Philippines) was to make it a policy for all the service stations to have clean toilets or latrines. I know it sounds elementary, but remember that the government was taking over an American-owned corporation. In those days, many people believed that anything American was far better than whatever Filipinos could come up with. Part of our marketing policy, then, was to ensure the maintenance of toilets in the service stations to show motorists that things did not become bad just because the company was now under Filipino ownership. I was very strict with that policy, and made it a point to make spot visits to various stations. I fired two sales agents from different stations for failing to comply. After that, the toilets were kept so clean you could sleep in them!

In the context of the oil crisis and PNOC's mandate, Petrophil's operations unavoidably took on the character of public service. Yes, my executive staff and I were committed to running the organization professionally and efficiently, and making the company profitable was our foremost goal. But being part of a government-owned corporation, we also kept in mind that whatever we did must contribute to the government's goals of securing oil supplies, reducing our dependence on oil-based products, and promoting national development. Unlike in a private firm where the CEO's principal responsibility is to keep the shareholders happy, in a government corporation like Petrophil our job was ultimately to promote the national interest.

For their part, the foreign oil companies looked with suspicion on PNOC and Petrophil. After all, the preferential crude oil that PNOC procured under government-to-government contracts was sold by Petron, and its freight costs were very competitive because government-owned tankers transported the crude. From the standpoint of our competitors, Petron was enjoying an unfair advantage because its operations were being supported by government resources. However, for as long as we kept to the old Filoil-Esso market share of 35-40 percent, the oil companies kept their peace.

But tensions between PNOC and the oil majors were unavoidable. Frankly, when the oil shock came, they were indifferent over the country's precarious oil supply and simply continued to sell crude in precrisis volumes. They did not cooperate at all when I first sought their help in obtaining government-to-government crude supply; I practically went door-to-door in the Middle East and they never helped. As PNOC built its capacity to obtain additional crude supply, my confidence in dealing with the oil majors also grew. Eventually, I required them to purchase 50 percent of their crude oil through government-to-government arrangements—this meant buying from PNOC rather than their commercial suppliers, which could affect their principals' net recoveries. The oil majors resisted this at first, but they relented because they could also not guarantee that they would be able to provide a steady crude supply for the country.

As PNOC began to flex its political muscle and tried to make the oil majors heed the government's crisis-management policies, the oil majors lobbied hard with President Marcos into restraining PNOC (or me, essentially). The head of Caltex, Ray Johnson, frequented Malacañang and harped to death about how I was curtailing private initiative and making decisions without consulting the majors. I went back and forth to the Palace to explain to the president what was going on. Fortunately, President Marcos understood and supported my position, although he advised me to accommodate the foreign oil companies without compromising PNOC's tasks.

I found a way to challenge and assuage the foreign oil companies' concerns when I decided to hire various senior executives of all the oil majors in the country, in the persons of Raul Paredes (Caltex), Bienvenido Gonzales Jr. (Mobil) and Antonio Mackay (Shell). I made them vice presidents of PNOC. They had equal access to company records and I had no problem with that. Although I knew they could provide information to our competitors, I treated them as bona fide management staff. After that, the oil majors could no longer claim that they had no opportunity to provide inputs on how government should deal with the oil crisis.

When the second oil crisis came in 1979, I was a little more secure and comfortable in coping with the situation compared to that in 1973. By then I knew the ropes and had access to friends in the Middle East. I also had an understanding of how the majors operated, their strengths and weaknesses, and so I was better prepared to handle them. This time around, I made sure the foreign oil companies would not be able to afford noncooperation the way they did in 1973. Since I had previous exposure to the past crisis, I was more confident in dealing with the vagaries of oil transactions in the world market.

Even our partnership with Mobil Oil in the operation of the Bataan Refining Corporation (BRC) was marked by occasional conflicts over the refinery's priorities. I mentioned in the first chapter how the American managers of BRC had refused to refine the crude oil that PNOC bought from China because of poor quality. They were unwilling to bend over backward to see what could be done, and it took all the persuasive powers of Jose U. Jovellanos, my assistant for Bataan Refining, to make the Americans accept the China crude. It was after that experience that I decided to hire the Americans as PNOC consultants, which required them to relinquish the refinery's management to us. Eventually, Mobil sold to us its 43 percent share, making BRC a wholly government-owned refinery.

Let me stress that I never thought of nationalizing the oil industry, and neither did the president nor the Cabinet. We always envisioned the industry to remain driven largely by free enterprise, with the private sector taking a lead role. Petrophil, despite being government-owned, had to compete on its own merits and abilities. I never aspired to make Petrophil monopolize the market. As I kept telling my management staff, if we became the only or the overwhelmingly large player, then the industry would become all-government and Petrophil would have no third-party test as to whether it was operating as efficiently as the private sector.

Of course, being a government corporation, Petrophil was not immune to political pressure on matters of product pricing. Yet, despite obvious pressures, we managed to walk the tightrope of fair market competition, and our pricing policy generally followed the movements of the international oil market. Because of our strong position vis-à-vis the foreign oil companies, Petron ended up dictating retail prices most of the time, to the benefit of consumers. We reflected true costs and still made profits without overpricing our products. At the same time, if we had to sell petroleum products at prices lower than costs, we eventually recouped the difference from subsequent operations.

Did the situation make the foreign oil companies uncomfortable? I am sure it did. But I want to emphasize that the Marcos government still left the foreign oil companies with a lot of room. They still had their market shares. They participated in government-initiated mechanisms to manage the oil industry. When an economic crisis struck us in 1982 and the country's international lenders compelled the government to undertake structural reforms, crude-oil procurement was deregulated and foreign oil companies were free to source all their supplies from the international spot market.

By the early Eighties, the Philippine economy was under serious threat. The domestic and external economic imbalances suffered in 1982 were, in significant measure, a function of depressed world prices for major commodity exports, reduced demand for nontraditional exports, burdensome oil import costs, protectionist policies by the major industrial nations that restricted access to important markets, and sharply rising interest payments on external debts following Mexico's default.

Our cost of servicing foreign obligations increased to US$ 2.23 billion, exacerbated by a 20-percent decline of foreign investment in the latter part of 1982. Consequently, we became even more dependent on short-term borrowing, which increased by 16 percent at the end of the year. In response, the International Monetary Fund and the World Bank imposed extensive reforms in our economy. The availability of loans was made subject not only to the introduction of a number of tariff reforms designed to open the domestic market to international competition, but also to more effective financial controls and an effort to increase the government's revenue-collecting measures.

The Aquino assassination in August 1983 pushed the economy over the hill, so to speak, producing perhaps the most profound crisis in postwar Philippine history. For ten days after the murder of Aquino, the archipelago was immersed in grief and talk about the faltering health of President Marcos circulated persistently. For years, he had been rumored to be suffering from *lupus erymatosus*, and it was said that his ailment made him incapable of discharging his duties.

Obviously, given the already precarious economic situation, the last thing President Marcos needed was a crisis like the Aquino assassination. Immediately after the event, there was instantaneous capital flight, an abrupt and wholesale calling in of short-term loans, and a suspension of new advances by international creditors apprehensive over the political situation. The interruption of short-term credit restricted access to critical productive imports and this, in turn, produced significant reductions in commodity output. The "cascade" effect included a drop in export earnings and an increase in the balance of payments deficit.

With the failure of international confidence in the country's stability, there was pressure to devalue the peso. In early October 1983, the peso devalued by a painful 28 percent—the second devaluation since June of the same year, when the peso was devalued by 9.4 percent. The prices of various commodities under government price control immediately rose by 15-30 percent, while prices outside the price control rose three to four times their predevaluation levels. By the end of 1983, the inflation rate had escalated to about 23 percent, and Manila requested a long-term rescheduling of 40 percent of the outstanding external debt estimated to be about US$ 24 billion at the time. The Central Bank was increasingly called upon to grant emergency funds to local financial institutions in order to provide the liquidity to cover mandated deposit-reserve requirements. Given the limited funds available, commercial banks raised interbank rates from 20 percent in 1983 to 33.45 percent in November, peaking briefly at 55 percent in December 1983.

The country suffered another blow with the onset of the worldwide Third World debt crisis in 1984, which raised interests on debt payments to record levels. As I said in the company's annual report for the year, it was a reversal of situation for PNOC, "from a crisis footing in the energy area under a relatively stable economy [in December 1973], to an economic crisis in a stable global energy scenario."[103] The restructuring program compelled all government corporations, including PNOC, to come under IMF-led periodic review; this significantly affected PNOC's capital spending, inventory policy, and management of receivables. The scarcity of foreign exchange forced the government to reduce PNOC's budget for oil imports to US$ 1.6 billion, a substantial decrease from the 1983 actual oil bill of US$ 2.1 billion. The situation also forced the Ministry of Energy to cut back on some energy-development projects and postpone the construction of several power plants.

Ironically, the financial crisis of 1984 proved how successful our energy development program had become. Although total national energy consumption in 1984 was substantially the same as in 1983, we managed to service the country's needs because of other energy resources like geothermal, coal and Palawan oil, which combined to generate 39.5 million barrels of fuel oil equivalent and amounted to a total displacement of imported oil worth about US$ 1.1 billion. As a result, we even saved on our 1984 budget of US$ 1.6 billion for oil imports by spending only US$ 1.45 billion.

I want to add that PNOC maintained a high credit standing among foreign lenders despite the country's dire financial position. We made use of this advantage to avail ourselves of short-term dollar credit lines with suppliers and foreign commercial banks to finance operations. In addition,

---

103. Philippine National Oil Company, *Annual Report 1984* (Makati, Rizal: Philippine National Oil Company), 2.

upon the request of Central Bank Governor Jaime Laya, we borrowed from foreign creditors and placed the dollar proceeds with major government financial institutions. I think it is fair to say that PNOC contributed significantly to government efforts at shoring up the country's reserves.[104]

It was also in this context of severe crisis that the government set up the Oil Price Stabilization Fund (OPSF) on October 15, 1984. Through Presidential Decree (PD) 1956, the government imposed an ad valorem tax on certain manufactured oils and other fuels, revised the existing specific tax on such products, and abolished the Oil Industry Special Fund (see the first chapter). Proceeds from the ad valorem tax were to be deposited to the General Fund, within which a Special Account called the Oil Price Stabilization Fund would be created "for the purpose of minimizing frequent price changes" brought on by fluctuations in exchange rate and oil prices. The OPSF would be used "to reimburse the oil companies for cost increases on crude oil and imported petroleum products resulting from exchange rate adjustment and/or increase in world market prices of crude oil."[105]

Before the creation of the OPSF, the procurement of crude oil for the country had already been deregulated, following the requirements of international lenders. Whereas previously the oil companies were required to buy half of their crude supply from PNOC, by 1983 the oil majors were free to source everything from the spot market. As a result, some oil companies were getting cheaper crude while others were getting expensive ones; because prices kept moving up or down, there was constant adjustment in domestic prices. Given the temper of the times, a one- to two-centavo increase already caused a lot of protest. In consultation with Board of Energy (formerly the Oil Industry Commission), which was setting the domestic oil prices, the Ministry of Energy suggested that all the industry players should sell at one price based on the average prices of all the three oil companies (Petron, Caltex, Shell).

---

[104]. See also Philippine National Oil Company, *Annual Report 1985* (Makati, Rizal: Philippine National Oil Company), 2-3.

[105]. Presidential Decree 1956, "Imposing an ad valorem tax on certain manufactured oils and other fuels, bunker fuel oil and diesel fuel oil; revising the rates of specific tax thereon; abolishing the Oil Industry Special Fund; and for other purposes," October 15, 1984.

How did the OPSF operate? The pump prices were set on the average, and the ministry monitored the landed cost of crude oil being brought in by each of the members of the oil industry. The ones with a lower average cost—the difference between true cost and the average pump price set by the Board of Energy—had to pay into the OPSF. On the other hand, those who had costs higher than the average set by the board withdrew from the fund. Through this mechanism, we managed to keep pump prices stable. When the average price generally declined, we just kept collecting OPSF contributions to ensure that we had enough buffer against future price spikes.

I want to emphasize that the Marcos government did not shell out a single centavo to finance the OPSF. Although PD 1956 stipulated that part of the revenues from various taxes on petroleum products would finance the OPSF, in reality, it was built through the contributions of the oil companies themselves. That is why when the Batasang Pambansa (National Assembly) wanted to appropriate a certain amount from the Fund to finance government expenditures, we explained to and advised the assemblymen that *they could not touch the OPSF since it was not owned by the government, but by the oil companies*. Government had no participation other than the fact that the Ministry of Energy managed it, i.e., monitoring the oil players' landed cost of crude, receiving contributions, and processing reimbursements. After a while, we renamed the OPSF as Oil Industry Equalization Fund, to reflect more accurately its nature as an oil industry fund used to equalize the members' costs.

Since the ministry was managing billions of pesos, we asked the Commission on Audit to examine all disbursements and collection. In addition, we published a status report on the OPSF every six months and gave copies not only to the oil companies but also to the president and the entire membership of the Batasang Pambansa. In this manner, we protected the ministry and the national government from accusations that we were pocketing the oil companies' money. I am proud of the fact that there was not one iota of rumor about corruption over the OPSF. Even the oil companies praised us highly for a job well done, and I want to give credit to Federico E. Puno, who conceptualized the mechanics and later took charge of managing the fund. He himself signed the checks, and I had minimal participation in administering the system.

I am emphasizing the private character of OPSF because it seems that, under President Marcos's successors, the OPSF either had been treated as a government fund or at least misconstrued as such. The OPSF was abolished by the enactment of RA 8479 ("Downstream Oil Industry Deregulation Act of 1998") on February 10, 1998; in subsequent years, however, the pressure to bring it back grew strong in the face of rapidly increasing oil prices. Sometime in May 2005, for example, the Philippine Chamber of Commerce proposed the OPSF's reinstitution although the Arroyo government's finance and energy officials rejected the idea since it would supposedly cost the government Php 13 billion (at Php 1 billion a month) to subsidize oil prices.[106] I am not aware of how the Aquino and Ramos administrations handled the OPSF, but the Marcos government certainly did not provide any subsidy to the OPSF.

By the end of 1985, my last year in office, PNOC's consolidated revenues had grown by 6 percent to Php 22 billion after contracting by 23 percent in 1984; much of the increase was from gains registered in geothermal and coal sales. Petron also remained the market leader in 1985 with a 38-percent share despite heightened competition. These and other factors increased PNOC's net income to a record Php 850 million, from Php 709 million in 1984. PNOC even declared a cash dividend of Php 240 million (of which Php 120 million was remitted to the national government as of January 9, 1986), and its tax payments made up 9 percent of government's total collection for 1985. Clearly, notwithstanding the deteriorating economic and political environment, and despite the constraints brought on by economic restructuring, PNOC showed its capacity to survive and remain profitable in a hostile environment.[107]

At the beginning of this chapter, I mentioned that the Ministry of Energy and especially PNOC became a terrain over which several Aquino

---

106. "Philippines to spend 13 bln pesos on fuel subsidy: official," *The Philippine Star*, March 31, 2005; see also "Finance department cool to PCCI idea of reviving oil fund," *Philippine Daily Inquirer*, April 3, 2005.
107. PNOC, *Annual Report 1985*, 3-7.

advisers fought to advance their agenda, foremost of which was the dismantling of the ministry and the privatization of PNOC, particularly Petron. Although they were not able to pursue Petron's privatization, Mrs. Aquino's subsequent moves in privatizing GOCCs can be seen as having paved the way for the more extensive privatization program of her successor, Fidel V. Ramos.

Nonetheless, it broke my heart when President Ramos decided to sell 40 percent of Petron to a subsidiary of the Saudi Arabia-owned Aramco. Before making the decision, he called for me to solicit my thoughts. I remember strongly advising against Petron's privatization, saying that it was the most profitable government corporation and thus a valuable source of revenues for government. More important, Petron performed a unique role in stabilizing local oil prices; from a policy perspective, Petron's strong market position was critical because it literally kept the foreign oil companies on their toes. Petron was the "crown jewel" of the government so why dispose of it? But President Ramos had other ideas. He felt that, precisely by selling the country's "crown jewel," he would prove his sincerity and determination in privatizing the economy. He chose to sell to Aramco because, as he reasoned with me then, it would guarantee the country's oil supply from Saudi Arabia.

In my opinion, after the mothballing of the nuclear plant, the privatization of Petron was the second-worst decision after Edsa I. I honestly think that not enough study went into President Ramos's decision: there was no evaluation, much less appreciation of Petron's intrinsic role as a potent policy instrument vis-à-vis the oil majors. At the same time, the company's value went beyond keeping the foreign oil companies in check. In my time, Petron was an independent source of revenue for PNOC and the Ministry of Energy; its profitability enabled us to finance various energy development projects without relying on government assistance. Petron's financial performance also made PNOC enjoy high credibility among international lenders, be they multilateral or private institutions.

Generally, I think the government should seriously consider before following the advice of the World Bank and other foreign creditors about privatizing government assets that perform a critical policy function. From our experience, privatization and deregulation have not really brought any substantial improvement in key economic sectors. If the intention was to break up monopolies and promote competition, I do not think anyone can say without batting an eyelash that this has been truly achieved under privatization and deregulation. The local oil industry itself is a good example.

Today, we are living under a regime of oil deregulation, and yet foreign oil companies still dominate the oil industry. Even Petron can be considered foreign-controlled despite the Philippine government's retention of 40-percent share in the company. Hence, the government can no longer use Petron as a policy instrument, especially in case of a severe oil crisis. In the face of steadily increasing oil prices, as what we currently have, the government cannot just order Petron to reduce its pump prices unless it manages to convince its counterpart from Aramco, which is highly unlikely.

Over the past year, as oil prices steadily rose to reach as high as US$ 70 per barrel and remain within the US$ 58-62 price range as of this writing, many quarters have called for a repeal of the oil deregulation law, the reinstitution of the OPSF, and even a government buy-back of Aramco's 40-percent share in Petron. These are unrealistic suggestions, in my opinion. Just how can government buy back Petron from Aramco? Not only will this be very expensive financially, but it can also invite retaliation. Think of how this would affect the status of the one million Filipino workers in Saudi Arabia and other Middle Eastern oil countries.

Unfortunately, conditions have changed drastically that reverting to the old setup prior to liberalization and privatization will be very difficult. Besides government's huge budget deficit, advocates of globalization—as represented by the country's international lenders—will likely meet with hostility any move by the government to reverse current liberal policies, especially in a strategic sector like the oil industry.

So what can government do? I will grant that the downstream oil market's deregulation has severely limited government's options in protecting consumers from pump-price increases. But I will also argue that government has been remiss in asserting itself vis-à-vis the "Big Three" (Shell, Caltex, Petron). The government still has regulatory and taxation powers that it should wield in order to discipline industry players that behave like a cartel in a supposedly deregulated market. Even the Philippine president possesses tremendous clout that should be wielded to call the oil companies' attention.

Over the longer term, and this would require legislation as well as strategic planning, I believe that we should move toward gradually but inevitably reducing the foreign oil companies' presence in the country.

Earlier in this chapter, I asserted that foreign oil companies have already made huge profits from their local operation. Although my case may be weakened by lack of data, even old ones that I no longer have access to, from my own experience at PNOC I can attest to the tremendous profits that can be made from oil retailing. Sadly, as in other economic sectors dominated by foreign corporations, the profits from oil retailing never stay in the country since these are remitted to mother companies headquartered in the US or Europe. To illustrate my point, I am referring you to a confidential, in-depth study on the market position of Getty Oil (Philippines) Inc. that PNOC made in 1974 (see the full text of my memorandum to President Marcos on pages 185 to 188).

At the time, all the foreign oil companies brought in their crude requirements except for Getty Oil (Philippines), which bought its supply requirement from Caltex (Philippines). We requested Getty Oil (Philippines) that they bring in four million barrels of crude oil, which was equivalent to their finished-product requirements for the year, but the company was unable to do so. I decided to commission an internal, confidential study on the company because I myself wanted to know what role exactly Getty Oil (Philippines) played in the local market.

To our consternation, we found out that it acted mainly as a trading company, with around 4.7 percent share of the market, and yet from 1969 to 1972, the company managed an average return of 28 percent on equity of Php 18.1 million. The company made annual remittances to their head office of Php 5.1 million as cash dividends, which was equivalent to a 100-percent dividend payout. Even at the time, the practice of 100-percent dividend payout was unusual; a normal corporation usually retains some of its earnings for further investment or for rainy days, so to speak. The remittances of Getty Oil (Philippines) not only showed how profitable the company was, but also the fact that they were not investing at all in the country. In fact, as the study showed, except for their service stations, Getty Oil (Philippines) did not make any capital investments in the country; it depended on Caltex (Philippines) for its supply and bulk plant requirements.

That is why I truly believe that, in a country like ours that does not have oil as a natural resource, locals should handle the marketing part of the oil business. In this way the profits, theoretically, do not slip out of the country. It can be done, as my team at PNOC and Petron had evidently shown. Today there are a number of Filipino capitalists endowed with sufficient capital to establish a marketing organization for oil. There is no secret formula in running an oil company, nor a sophisticated demand for technology; it is really a straight and simple marketing effort. That is why I submit that locals should handle the marketing. Of course, localizing the oil retail market will not eliminate the profit motive, and it should not be eliminated. But at least, theoretically, the money will not slip out of the country, and hopefully we have enough Filipino conscience to realize what margin is enough without imposing too much on consumers. This is how PNOC and Petron played the game in my time.

Oil is very sensitive to our national interest, to our daily expense—almost as important as fish and meat, and maybe even rice. It is for this very reason that we have to seriously consider that these businesses be nationalized in the sense that it is *Filipino-owned or Filipino-controlled*, not in the sense of government owning it since that will definitely discourage foreign business. I do not think that foreign companies should be kept out of the oil industry, because they still have a role to play in promoting research and development in other products like lubricating oil. The foreign oil companies have direct access to engine manufacturers; they work with them, introduce their oil, experiment with engine manufacturers on the use of their oil right in the engine factories. If it works out, the engine manufacturers can recommend the product for standard use with their engines. Since we do not have that kind of access to engine manufacturers, we will still have to buy lubricating oil and grease from the oil companies. But that is minor compared to the more important products like gasoline or kerosene.

I do not think there is anything so difficult in managing a local oil company. We can have access to product upgrades and information updates. With the Internet and other advanced communication technologies, we can access new knowledge and apply it here, which would invariably translate into lower production costs. Of course, whoever is running the Filipino-controlled oil company must keep abreast of developments abroad. We have to tap international research teams or collaborate with foreign companies in research and development, since we cannot do it ourselves.

Oil refining—from the design of oil refinery to actual processing—is actually the only major area of the oil business where foreign participation is necessary, since it requires engineering sophistication and huge capital. The refinery processes crude oil to give you various types of gasoline, various types of diesel and fuel oil, kerosene and, technically, aviation fuel. Oil refining is dominated by the Americans, although there are refineries owned by the English, Dutch, French, and Italian companies.

Right now, there are small companies in the market like Flying V or Seaoil that are operated by Filipinos. They buy from the cheapest sources, add a little margin and then they are able to distribute the product. Often, these companies also source their products from the oil majors; in a sense, you can say that this business arrangement with the smaller retail companies enable the oil majors to have additional outlets for excess inventory, sold at a slightly higher mark-up. You can actually make the case that smaller companies serve a purpose for the majors similar to the relationship between a wholesaler and a retailer. Especially when oil prices are high, the small oil companies manage to provide a "psychological shield" that the oil majors are pricing their oil correctly. That is why you really cannot say that there is competition in the oil retail market, despite the presence of these smaller companies. At the same time, we must remember that a Filipino company needs huge amounts of capital if it wants to compete respectably with the foreign oil companies. The problem with the oil business is that it is highly capital-intensive, so a Filipino oil company must be a substantial organization. It can be done if our leading businessmen just decide to pool their resources.

The message I am trying to drive home is that we have to start looking at the oil industry in terms of the national interest. As I said, oil is not a simple commodity that can be left to the vagaries of the market; almost all aspects of our daily life depend on oil—our transport system, the clothes we wear, the lifestyle we have. Even the lowliest Filipino needs kerosene.

We can begin by legislating a limit to the number of years that foreign oil companies can operate in the country, otherwise they will never leave as they have never left over the past hundred years. Sure, I will grant that these companies brought in capital when they set up operations here. But I will also argue that they have since stopped contributing capital investments once their infrastructure for refining, distribution, and marketing had been established—and this happened way back in the Fifties and Sixties. If I may say so, the oil companies have been living off the fat of the land primarily through their stranglehold of the retail market.

The oil companies will say that they are modernizing the distribution system, as evidenced, for example, by the service stations being built along the north and south diversion roads, making their stations on a par with those in the advanced world in terms of providing diversified services. But it is really just a marketing technique to sell their products. Is this the kind of capital investment we want from the oil companies? I would prefer capital investments in advanced refining technology and outlying bulk plants in order to pioneer development especially in far-flung areas.

Another reason why I advocate putting a specific timeframe to the operation of foreign oil companies in the country—say, another twenty years—is to force government to snap out of its complacency and formulate a comprehensive plan that would prepare the economy and local businessmen for the foreign companies' pullout. During the interim, the government can provide the necessary support to build local groups' capacities in managing an oil company.

From a policy standpoint, setting the stage for the phaseout of foreign oil companies in the local retail market would have been easier had Petron remained under government ownership. When I read in the papers that

Petron had a total net income of Php 5.73 billion from 2004 to the first semester of 2005, I could not help but feel a sense of loss. Petron had always generated profits, and its strong market position was evident in its domestic net income, which totaled Php 3.4 billion for the reported period.[108] Assuming that Petron still holds 40 percent of the local oil market, we are talking of about Php 8.5 billion in domestic profits earned mostly by the foreign oil companies.

Speaking of Petron's profits, I have been asked by a number of government employees who had bought Petron shares during its initial public offering (IPO) in 1994, whether it was a worthwhile investment for them. Although I have consistently stated that Petron has been making tremendous profits, I think its IPO selling price of Php 9 per share was too high since, in relation to its present value in the stock exchange of roughly Php 4 per share and earnings per share of Php 0.37, this initial offering of Php 9 per share translates into a selling price of roughly twenty-four times present earnings, which is too high and unrealistic. The cash dividends paid out at Php 0.15 per share in 2002 and Php 0.20 for 2003 and 2004, respectively, were definitely more than average by Philippine Stock Exchange standards. If the same cash dividends were paid on the basis of the current selling price of Php 4 per share as of this writing, definitely investing at Php 9 per share was much too high when Petron was publicly offered in 1994, and is difficult to justify under present operating incomes. Petron's profits should be at least twice its current profits to give good reason for the initil investment of Php 9 per share.

But what arouses my curiosity about the Petron shares is, how much did Aramco really pay for its 40-percent holdings in the company? I have received unverified reports that Aramco paid much less than Php 9 per share. If this is true, then it raises the ticklish question of why private domestic investors should pay more than a foreign oil company. The government should clarify this matter for the sake of the private shareholders, who are mostly former Petron and government employees.

---

108. "Petron shells out P100M yearly for PUV discount," *The Philippine Star*, September 1, 2005.

On the other hand, the private shareholders who own 20 percent of Petron should also take it upon themselves to raise such issues. Perhaps they should consolidate themselves and insist on naming their representatives to the board of directors which, as currently constituted, has ten seats. They should exercise their right to be represented by two board members of their choice so that their interests could be properly represented.

From a broader perspective, the report on Petron's profits convinced me even more that we should seriously think about why we continue to accommodate these foreign oil companies. For more than a hundred years now, we have made them profitable without compelling them to make commensurate investments in the country (although it has been reported that Aramco has been investing, perhaps in the form of loan accommodations to Petron for the enhancement of its refinery operations). Take the case of Pilipinas Shell, which was reported as seriously considering the shutting down of its Batangas refinery, which has a capacity of 153,000 barrels per day, due to "weak refining margins"; Caltex (Philippines) had earlier closed its Batangas refinery for the same reason. Pilipinas Shell is demanding that the government impose higher tariffs on imported refined products if it is to convince the parent company Royal Dutch Shell, to continue operating the Batangas refinery.[109]

Personally, I find Shell's complaint over "weak refining margins" flimsy. Petron also has a refinery and yet the company made huge profits domestically and otherwise, since 38 percent of the company's net income came from the export of refined products. I do not think this looks like "weak refining margins." If oil companies like Shell and Caltex prefer to import refined fuel products for domestic sale, then that makes them a mere trading company, as Getty Oil was in the Seventies. If the oil companies do not want to put up the infrastructure for oil processing and distribution, then it gives us more reason to question the role of foreign oil companies. For too long, we have been very accommodating to them. I think it is time to ask ourselves, What have these companies really contributed to our country to deserve such generous accommodations?

---

109. "Shell says ball in gov't court on questions about refinery," *BusinessWorld,* September 2-3, 2005.

GZV's Memorandum to President Ferdinand E. Marcos

January 25, 1974

Memorandum to: His Excellency
                 President Ferdinand E. Marcos
                 Malacañang, Manila

Re: Getty Oil (Philippines) Inc.

In the light of Getty's reported inability to bring in the crude equivalent of their finished product requirements, we made a screening-type study on Getty's role in the Philippine economy. Our findings are as follows:

1. Getty Oil (Philippines), Inc. is mainly a trading company with approximately 4.7 percent of the Philippine oil market. Their total assets as of yearend 1972 are Php 38.7 million, with stockholders' equity of Php 18.1 million. They have 557 retail outlets (234 in Luzon and 343 in the Southern Islands) of which 80 percent are dealer-owned. Their main products are mogas (50 percent), diesel (33 percent) and lubes/greases (4 percent). They have 274 employees.

2. Getty from 1969 to 1972 derived an annual net income of Php 5.1 million, a 28 percent return on the Php 18.1 million equity. They remitted to their head office Php 5.1 million annually as cash dividends, a 100 percent dividend payout.

3. Getty has not made any significant investments in productive facilities as they depend on Caltex for their supply and bulk plant requirements. Their investments are primarily in service stations.

4. Getty does not appear to have any specific knowledge or technical skill not available to other members of the industry.

We conclude, therefore, that Getty's economic contribution to the country is minimal and a phase-out of Getty's Philippine operations would cause no significant dislocation to the Philippine economy.

Although a phase-out may have political repercussions vis-à-vis our U.S. relationship, the type of foreign investor the Philippines desires is one which makes a positive contribution to the present development efforts. Without providing crude supply, their presence here is anachronistic, and Getty may have outlived its usefulness.

I suggest we adopt the following strategy:

(a) Allow Getty a time period within which to bring in their own crude (4 million barrels per year) say until June 30;

(b) Should they fail to do so, notify Getty that their Philippine operations will have to be curtailed by yearend-1974.

The above is a suggestion, and we would appreciate receiving the President's comments.

ALC:GZV:aal

G.Z. Velasco

Attach.

GETTY OIL (PHILIPPINES) INC.

1. **Financial Structure, MP**

	1972	1971	1970	1969
Current Assets	<u>26.3</u>	<u>23.9</u>	<u>21.3</u>	<u>17.3</u>
Receivables (Net)	10.3	10.0	9.2	6.7
Inventory	7.9	7.1	4.3	7.2
Others	8.1	6.8	7.8	3.4
Property, Plant & Equipment (Net)	10.7	11.2	11.7	10.7
Others	1.7	1.0	0.8	.4
Total Assets	38.7	36.1	33.9	28.4
	===	===	===	===
Current Liabilities	<u>20.6</u>	<u>17.7</u>	<u>15.7</u>	<u>10.8</u>
Accounts Payable	12.1	10.1	9.6	6.1
Due to Getty Oil – Phil. Branch	4.4	2.5	1.2	0.4
Others	4.1	5.1	4.9	4.4
Stockholders' Equity	<u>18.1</u>	<u>18.4</u>	<u>18.1</u>	<u>17.6</u>
Capital Stock	14.6	14.6	14.6	14.6
Retained Earnings	3.5	3.8	3.5	3.0
Total	38.7	36.1	33.9	28.4
	===	===	===	===

2. **Income Statement, MP**

	1972	1971	1970	1969
Sales	119.5	109.5	91.1	71.7
Cost of Sales	97.5	85.2	68.4	47.2
Gross Profit	22.0	24.3	22.7	24.5
Operating Expense	18.7	17.0	15.6	13.4
Income from Operations	3.3	7.3	7.1	11.2
Other Income (Charges)	0.7	.8	.5	.4
Income Before Tax	4.0	8.1	7.6	11.6
Net Income	2.6	5.3	5.0	7.5
	===	===	===	===
Cash Dividends	2.9	5.0	4.5	8.0

3. **No. of Retail Outlets, 1972**

	<u>SID</u>	<u>Luzon</u>	<u>Philippines</u>
Company-Owned	52	60	112
Dealer Owned	291	174	465
Total	343	234	577

4. Sales Volume by Product, MB   1972   1973 (9 Months)

        Mogas   1297.9   1090.9
        Kero   280.3   228.8
        ADO   774.5   680.6
        IDO   157.4   104.4
        IFO   161.1   131.4
        Lubes   117.8   116.8
        Greases   2.6   2.0
                        ――――――   ――――――
        Total   2791.6   2354.9

5. Sales Volume by Business Line, MB   1972   1973 (9 Months)

        Retail   1878.7   1631.9
        Consumer   751.4   607.3
        PI Government   107.6   95.9
                        ――――――   ――――――
        Others   53.9   19.8
                        ――――――   ――――――
                        2791.0   2354.9

6. Getty Oil (Philippines) Inc. lifts its products from Caltex supposedly at prices based on incremental costing and therefore even lower than the prices Caltex bills its own dealers. For its lubes, Getty (LA) has a supply arrangement with Exxon for which Exxon pays a royalty of 3 cents/US gallon. The basestocks are delivered from Singapore and the U.S. at about the same price as that for Petrophil. In turn, Getty Oil (Philippines) sells to Filoil and Theo H. Davies, by virtue of a long historical relationship. Getty Oil (Philippines) is not an original shareholder of PPC but is believed to have bought some PPC shares in the open market. It is also supposed to have a lube supply contract with PPC.

VCA:aal
1-23-74

Note: Some figures were rounded off and do not tally.

Postscript

For me, government service was a rewarding experience. It enabled me to reach what psychologists call "self-actualization," and the biggest reward I had was the fact that I was able to work very well with about thirty key executives and staff who were professional and dedicated in meeting the challenges that lay before us when we started PNOC over three decades ago. I am proud of the fact that not one of the original group of executives from the old Esso (Philippines) left PNOC. After the takeover, I remember talking and pleading with each one not to leave, to try out my system for six months, and if it did not work out, they were free to go. Not one of them did, and together we met the oil crisis head-on and built the country's energy infrastructure from bottom up for the next twelve years.

Besides my management experience I think another reason why I was able to exercise my leadership and direct these executives toward our mission was the fact that I was not a politician, and it reflected in the way I handled my job and dealt with the people I worked with. They saw that I had no political agenda, and whatever I asked them to do was consistent with corporate objectives rather than for personal or political gain. I even prohibited my wife from performing ribbon-cutting functions; it was only in the later years that I allowed her to occasionally cut ribbons.

Taking on PNOC, and later the Ministry of Energy, was the most challenging job that an engineer could ever have. The scope and scale of our efforts were far beyond what I had dreamed of doing when I was just an engineering student. It was a dream for any engineer to do all those projects that we did; although I did not do the engineering myself, managing the entire system necessitated that I had an engineer's comprehension and appreciation for the different projects that we carried out. I doubt that anybody else was given this kind of opportunity. Of course, I do not deny that the "perquisites" that went with the job were part of the attraction. I had transport means that enabled me to visit project sites; President Marcos told me to get a Lear Jet to facilitate my various travels domestically and to other countries, especially the Middle East where air transport facilities were still limited.

In sharing our experiences in developing the country's energy program, I tried to show you how, from a crisis situation following the oil shock of '73-'74, the government struggled to build its capacity to address the country's energy needs. Our response came in the form of an energy-development program anchored on reducing our dependence on imported oil through the diversification of our energy sources. Our battlecry then was "energy self-reliance." It sounds easy today, but at that time, my team and I had to do a lot of studying before we could even start the work.

To be able to develop an energy policy, you have to have an inventory of the energy sources available to you—oil, coal, geothermal, hydro, etc. When we were tasked to formulate an energy policy, we discovered that there was no inventory available, and we had no idea where to obtain the information. What were the resources that we could use? Where were these resources? We did not know.

When the oil crisis began, the country was 97-percent dependent on oil for energy, and the remainder was largely provided by hydro (with minimal contribution by marsh gas and animal waste). However, once we set our ten-year energy policy in 1975, we gradually established the possibility that we could reduce by half our dependence on oil and replace it with non-oil sources—hydro, geothermal, coal, nuclear, alcogas, and others. We had targeted a 50-percent reduction in oil dependence by 1985; we reached 46 percent as 1986 began. For us at PNOC, diversification of energy sources and reducing oil dependence were crucial if we were to minimize the country's vulnerability to rising oil prices.

Unfortunately, when toward the latter part of the Eighties, world oil prices fell to as low as US$ 10 per barrel, the Aquino administration rationalized its complacency by citing the world price situation as a justification for dismantling our energy program. But what about today, when crude continues to fetch more than US$ 60 per barrel? The singular lesson to be learned from the '73-'74 oil shock is that you will never know when the "era of cheap oil" will end, and therefore government must be prepared for contingencies. That was the whole point about our energy development program: We could not afford to be shortsighted and miserly in investing resources—like filling up your car with just enough gasoline to get you to your immediate destination. No, in the case of energy development, we were always driven by the thought that another oil crisis could break out anytime (as in fact it did, in 1979).

I believe that my experiences in running PNOC and the Ministry of Energy yield important lessons that are worth sharing with the present generation of managers in government. It is true that the political environment I worked in was completely different from what we have now. There was rigidity in the system then, in the form of centralized decision making, which is thoroughly the opposite of what we have today. But I still think that, faced with a situation similar to what I confronted in the early Seventies, today's government manager can adapt according to the present political framework, although he has to operate under certain constraints that would prevent him from being as decisive as we were in the past. Back then, we were driven by a single objective—to reduce our country's dependence on imported oil—and it did not matter if we had to shortcut the process (i.e., doing away with government audit or constraints in hiring) just to attain our objective. This was made possible by the martial law environment.

In other words, I have to admit that the conditions at the time made it easier to institute development work similar to energy. Our work at PNOC entailed going into essentially new development areas that were totally foreign to us. The organization was geared for oil refining, marketing, and distribution, but we built on these primary characteristics to expand our expertise to other energy-related sources. In our quest for self-reliance, no energy form was insignificant enough not to be tried out. As you have seen in the preceding pages of this book, what we accomplished was the creation of a comprehensive energy infrastructure whose components, with the exception of hydro power and a little of geothermal, were built from virtually nothing.

But we could not have done it were it not for the availability of corporate, scientific and engineering expertise. Besides the key executives whom I inherited from Esso (Philippines), we managed to attract executives from other oil companies. Equally important, scientists and technologists from the University of the Philippines were enticed to join us, encouraged by the fact that we were doing the kind of work that enabled them to contribute as well as enhance their knowledge. The sophistication with which we were able to implement the various energy projects I really owe to the expertise of the people that I have been fortunate to lead.

When I look back at those days, I cannot help but marvel at the fact that we never encountered any major obstacle in transforming PNOC from a national oil company to a "total energy company." PNOC made the Philippines the world's second largest producer of geothermal energy, it rationalized the use of coal as an energy source, it experimented with alcogas, coco-diesel and other indigenous energy sources, and it made Petron the biggest corporation in the country in terms of sales revenue. To this day, too, PNOC remains the only Filipino-owned corporation that ever landed in *Fortune Magazine's* Top 500 companies outside the United States, and we achieved this status for four consecutive years (1978-1981). This probably proves that a government corporation can be run efficiently and profitably. Another feather in our cap was the World Bank's citation of our energy program as a "model for Third World countries."

To my mind, the most crucial ingredient to our success was that we organized subsidiary companies that would focus on each of the different energy fields. I believe that this management approach enabled the key executives to concentrate on their respective tasks without having to worry about how to finance or staff their activities because PNOC, as the mother company, serviced these needs. The head of each subsidiary had to present his plans and proposed budget to the PNOC executive committee (composed of the executive vice presidents and senior vice presidents) for approval. Once the proposal was approved, the Execom assured the head of the subsidiary that his needs would be met adequately.

I will also have to say that PNOC's insulation from politics contributed immensely to its sustained success. Hard as it may seem for his critics to believe, President Marcos vested so much confidence in us and gave me a free hand in running PNOC. For this, I came to admire and respect him even more. Because of the president's trust, I was able to shield PNOC and my executive staff from political pressure. We ran the company as if it were a private-sector business, and we made our decisions based on company objectives and bottom line. For all the accusations of corruption against us, to this day I can face anyone squarely and say that PNOC's money was never used for political and private purposes. Yes, there were occasions when we had to compromise with certain political realities, but these did not jeopardize PNOC's performance in any way.

As for external relations, an important factor in securing our country's crude supply during the first oil shock was the fact that I was able to develop friendships with several oil ministers in the Arab world. This was very important in cementing the Philippines' relationship with major oil-producing countries. I could pick up the phone and talk to the most powerful oil minister as though he were a long-lost friend. My investment in such relationships certainly contributed in ensuring that the country had a steady commitment of preferential crude supply from the Middle East.

I can proudly say that PNOC built the country's energy infrastructure, and we did this in the midst of an energy crisis. Unfortunately, Mrs. Aquino dismantled everything that my colleagues and I built, for the simple reason that we were "tainted" by our association with the deposed president. Even worse, the deregulation and privatization of the oil and power sectors by subsequent administrations have resulted in high energy costs, a situation that PNOC and the Ministry of Energy had precisely worked to mitigate, if not completely avoid, through our energy development program.

When we speak of energy costs, we essentially mean costs of energy from oil and its derivatives like gasoline, diesel, kerosene, jet fuel, aviation gasoline, and asphalt. In the case of electricity, the cost is normally measured at the final price for which a consumer is billed, be it commercial/industrial, or strictly consumer.

The pricing structure for oil is relatively simple. The basic cost component is the price of crude as purchased from its supplier, then you add the expenses incurred in transporting it to the Philippines and processing it in the refineries. Usually, it is in the refinery "fuel and loss" that one encounters significant cost differentials among variously sourced crude, since certain refineries are more efficient than others and conversely, some crude have less losses in refining than other kinds of crude.

What I find unusual about the present pricing structure is that, although each oil company operating in the Philippines has its own sources of crude and system of importation and costing, somehow the resulting market prices end up nearly identical. The three foreign players in the country today are Shell, Caltex, and Total, which obtain crude from various producing countries as they have worldwide access. It is quite difficult to rationalize how these oil companies come up with uniform market prices. No wonder that the oil majors invite suspicion because, frankly, unless they get together or literally connive, prices will not and cannot be the same for all of them.

In the past, when Petron was a wholly owned government subsidiary that had 40 percent of the market, the company determined its own price and the oil majors tended to follow. The consumers trusted Petron and felt assured that they were getting the right prices. Under present circumstances, however, with Petron effectively controlled and partially owned by Aramco, can the public still rely on Petron to price its products right? I do not mean to "cry over spilled milk," but if only Petron continued to be in government hands, we may at least have a reason to feel a sense of comfort every time we line up at the service station.

Another area of concern is the timing of price increases. In the past, as a matter of government policy, oil companies were required to maintain a minimum of sixty days' inventory to serve as cushion against market price increases or even decreases. This meant that local price adjustment to conform to an increase or decrease in the world market did not have to be implemented immediately, as the oil companies had a sixty days' cushion before they could apply any price increase/decrease. Right now, however, the spate of pump price adjustments made by oil companies every time world crude prices rise makes one ask whether they are being transparent in their pricing policies. We are left hoping that they are charging the right prices, unlike in the past when we had Petron to lead the final pricing structure.

Don't get me wrong. Pegging the "right" market price for oil is really a very difficult judgment call. Many factors must be considered and you cannot even compare prices with, say, the ASEAN region because each country has its own importation and taxation policies that can affect the final price. But what is noteworthy is that, as I had seen in the past, every time there was a price increase, the oil companies declared a higher profit for that particular year. In fact, this is now a big question facing the United States Congress given that, for instance, Exxon/Mobil is declaring very high profits. Congressional leaders are tempted to impose a tax on excess profit or windfall gains. The oil companies, however, argue that as world oil prices rise, exploration costs also increase substantially.

As for electricity, power generation in the past rested primarily in the National Power Corporation (NPC or Napocor), whose role was to generate and sell electricity to franchise holders that then distribute it to various parts of the country. Under the Marcos administration, the generation of electricity was limited to NPC, with the exception of some industries that would rather depend on their own generating facilities for their exclusive use. The rationale for such a scheme was the fact that Napocor should install high generating capabilities, so that the cost per kilowatt hour installed would be much less, and its resulting price also less.

The breadth and depth of NPC's mandate and organizational capabilities for power generation and transmission were conducive to its exploration of non-oil sources such as geothermal, hydro, and coal. As a rule of thumb, the higher the cost of installation, the less the generating cost per unit. Take the case of hydro—its cost is very high compared with other sources of energy, but its cost of generating electricity is much less, as only water is being used to drive the turbines. Essentially, the hydro system is a multipurpose system since, after the water is used by the turbines, it is flowed for the use of irrigation and even flood-control purposes. In the case of geothermal, the steam is tapped from the earth and is recycled back to its source in order to avoid any possible contamination of the ground or surface water system. These two types of energy source are not being used by private-utility franchises because no independent power producer could afford the huge capital requirements of a geothermal or hydroelectric power plant. The non-oil source currently attracting the interest of investors is natural gas, such as the one in Malampaya. Despite the huge cost of tapping and piping the gas from Palawan to Batangas, the Malampaya natural gas plant still comes out as more economical than oil-based thermal plants, although the cost of transmission is quite high.

It is sad that the past setup of maintaining NPC primarily for power generation is no longer possible, as it is now a nonviable organization due to past administrations' policy of pegging electricity rates below production cost, but especially oil price increases. Because of political considerations, Napocor did not collect the right amount of rate increases commensurate to the oil price increases and thus suffered underrecoveries in almost all administrations since President Marcos.

Ironically, Napocor's bankruptcy provided the justification for policymakers to devise a new system, courtesy of Republic Act 9136 ("Electric Power Industry Reform Act of 2001" or EPIRA), which allows franchise holders of electricity to generate power as well.[110] The problem is, while EPIRA exposed NPC to competition from its distribution customers, it could not sell power in the distribution companies' franchise areas; the result was significant excess capacity for National Power. Obviously, this is not an even playing field; what we have instead is a case of NPC competing with one hand tied behind its back.

Under EPIRA, distribution utilities should make their profits in distribution and not in the sale of electricity, which means to say that a utility firm cannot add an incremental margin of its purchased

---

110. The Aquino administration actually initiated the liberalization of power generation, when it issued Executive Order 215 dated July 10, 1987 that effectively rescinded PD 40, which had made the generation of electricity an exclusive function of Napocor.

electricity. Therefore, their objective should be to source for the cheapest electricity supply and not to generate it by themselves or internally. However, for purchase contracts already existing before EPIRA was passed, utility firms were allowed to charge the contract rate for the cost of purchased power even if it is higher than the market rate.[111] This is the case with Meralco's contracts with First Gen for the Santa Rita and San Lorenzo power plants.

As I have stated previously, as a matter of principle, I am firmly opposed to allowing franchise holders such as Meralco, Cepalco, Davao Light and Power, Visayan Electric Company and the like, to generate their own electricity and then turn around and sell the same to its franchise network. As I have said, this is a formula for conflict of interest: The distributor will always opt to buy from related power-generating companies at the highest possible price, or at least encourage them to match the lowest offers from other power generators.

Unfortunately, we face a *fait accompli* because the primary source of generation in the past, National Power, is now no longer capable of fulfilling such function and is even being phased out. The government should have maintained NPC as the primary source of generated electricity for sale to the respective franchise holders. Not only is such a setup suitable for economies of scale, but it also enables Napocor to lead the way in utilizing non-oil resources such as hydro and geothermal, whose high installation costs are beyond the practical reach of individual private producers.

But we can still remedy the situation, and we can start by amending EPIRA so that its mandate of ultimately providing low-cost electricity to consumers can be realized. We must also remember that energy costs are a crucial factor in investment decisions. In light of globalization, unless we reduce our electricity costs to a level that is competitive, we will suffer disadvantages in the international market. As it is, we are already handicapped by the fact that we do not have oil resources, and although we have hydro and geothermal as alternative fuel, these cannot be tapped by the franchise holders due to prohibitive installation costs.

---

111. Republic Act 9136, "Electric Power Industry Reform Act of 2001," June 8, 2001.

This brings me to the broader issue of privatization of the energy sector. Past and present administrations since the Marcos era have unquestioningly embraced the standard prescription of the International Monetary Fund, the World Bank and the Asian Development Bank that privatization be adopted as a matter of policy. I am particularly concerned about the fact that the PNOC Energy Development Corporation (PNOC-EDC) is being advertised for sale to private interests.

I submit that PNOC-EDC, the developer of the geothermal units currently installed in Tongonan, Leyte, and Palinpinon, Negros Oriental, is a showcase of consistently reliable performance and guaranteed financial viability due to its numerous take or pay contracts with NPC. PNOC-EDC initially developed the geothermal fields of Tongonan, Palinpinon, and Bacon-Manito, and Napocor constructed the geothermal power plants. Both companies subsequently entered into a Steam Supply Agreement for PNOC-EDC to supply Napocor's power plants for twenty-five years on a "take or pay basis," which essentially means that the latter would pay for the contracted volume of steam whether it be used or not. The present structure of Leyte "A & B" (which shares with Tongonan the same geothermal reserves), with a capacity of 600 megawatts, was awarded on the basis of PNOC-EDC developing the field and contracting California Energy, a US company, to construct, operate and maintain a power plant to convert the steam to electricity, which PNOC-EDC subsequently sells to NPC on a "take or pay basis." Similar arrangements were entered into with Marubeni, with reference to the Mount Apo 100 megawatts geothermal field. With these contracts alone, I do not see any justification why PNOC-EDC should be privatized. You are effectively selling a "cash cow"!

I foresee with certainty that should PNOC-EDC be privatized, only foreign companies will qualify as bidders because of huge capital requirements (about US$ 100-250 million). What I cannot understand is why we insist on privatizing government assets even in areas where the government has proven that it can run certain companies efficiently and profitably. Have we not learned our lesson from Petron? I opposed and continue to oppose the privatization of Petron, which fulfilled a function of national interest and was very profitable. PNOC-EDC is another shining example that privatization should be stopped where government can successfully run it; besides, given our need for energy self-reliance, the maintenance and continued success of PNOC-EDC should be of high national interest.

Addressing the high cost of energy by remedying flaws in the present pricing structures is just one aspect of a comprehensive energy policy that the Philippine government must craft, if we are to adequately confront new challenges as well as maximize opportunities offered by technological advances in the energy sector.

Generally speaking, having an energy policy such as the one we had is much more critical today, given the fact that oil prices have been rising steadily, reaching as high as US$ 70 a barrel as of this writing. As far as I know, this development is a significant departure from what industry observers and energy experts expect. From what I have read over the years since I left government, many energy experts did not expect a major change in oil policy until 2015, which is when the major oil-producing countries—Saudi Arabia, other Middle East countries, and Russia—are expected to reduce output because of the depletion of supply. Of course, these projections are based on estimates as well as known and proven reserves at a given time.

However, since China entered the picture with its insatiable demand for oil to power its growth, many energy experts have revised their projections about world oil supply. The recent failed attempt by the China National Offshore Oil Corp. (CNOOC) to outbid Chevron Corporation[112] in buying Union Oil Company of California (Unocal) speaks volumes not only of China's economic muscle but its determination to secure strategic supplies of oil. After all, Unocal—the ninth-largest US oil company—owned large oil-and-gas assets in Asia, not to mention geothermal exploration rights in the Philippines. The hostile response of American legislators and regulators to CNOOC's bid is equally instructive: when it comes to oil and energy issues, selling Unocal to a foreign group like the Chinese is a matter of national security.[113] India is another player that is

---

112. Formerly known as ChevronTexaco, it is the parent company of Caltex.
113. Unocal's stockholders eventually approved a merger agreement with Chevron Corporation in August 2005 that made Unocal a wholly-owned subsidiary of the latter.

expected to impact substantially on the world's oil supply, as the country's economic and political standing continues to improve vis-à-vis the rest of Asia. Between China and India alone, we are talking about a third of the world's population, although it is not so much the number of people per se but the improved buying power of a substantial population that can put pressure on oil supply. Developing countries like the Philippines, much of Africa and others, will inevitably suffer from such a situation because there is no way we can compete with China or India in buying an increasingly expensive oil.

I am sorry to say that nobody in government is thinking about the situation, much less worry over the world energy situation's adverse effects on our country. Somebody has to monitor this now. Of course, there are many sources of information, like the UN and the OPEC itself. But it is not what you need in order to be comfortable. We should monitor the growth in per capita income and corresponding oil consumption of other countries. Let us say one African country increases its per capita income—South Africa is a good example, being the biggest and most progressive country in the continent. Every increase in South Africa's per capita income will lead to a significant increase in its oil consumption. From what I remember, for every one percent increase in per capita income, there is a 1.5-percent increase in energy consumption. Assuming that the ratio still holds, we can use this to plot the oil consumption of particular countries and make projections on how it would affect oil supply.

In the Philippines, if we use the 2004 figure of 126 million barrels of oil that the country imported, this translates to a per-capita consumption of about 1.5 barrels a year, at an estimated population of 84 million. If these figures are correct, this per capita consumption of 1.5 barrels is the low end among world oil consumers; at the high end of oil consumers is the US, with a per-capita consumption of 60 barrels annually. What surprises me about our estimated per-capita consumption of 1.5 barrels of oil is that, it does not speak well of the quality of life of Filipinos. In my time, we already estimated that each Filipino was already consuming about two barrels a year. Even with population growth, oil consumption should rise with economic growth. If present per-capita consumption figures are accurate, these seem to indicate that the quality of life of most Filipinos is even deteriorating.

Anyway, my point is that somebody in government has to take care of studying and monitoring all this. You cannot leave it simply with the energy secretary. As constituted today, the Department of Energy has shrunk in its responsibilities. The National Power Corporation is effectively beyond the department's reach, and is to be totally privatized, as today it is being bid out. With Petron out of the picture, and with the PNOC Energy Development Company operating by itself and eventually to be privatized, what is left for the Department of Energy to do? To my mind, it is simply to make sure that there is enough oil available in the country at all times. Even the matter of price determination for electricity is outside the realm of the department, which is structurally correct because the regulation of power rates does not fall within the principal functions of the department. And with the oil deregulation law, the department has even less influence in the setting of pump prices.

Given the situation, my suggestion is that the energy department be downgraded to a commission and placed directly under the Office of the President. As a commission, it will still undertake a crucial role in directing our energy policies, since its primary function would be to review the energy situation internationally and domestically, to make sure that the policies of government conform to what is expected to develop in the immediate future. This is actually about the only function of the energy department that has not been transferred, and to me it is a sensitive function, that is why it is important to professionalize this organization and ensure that members cannot be removed by politics or changes in administration. Since the country is heavily dependent on oil, this professionalized planning group will be the one to sound the alarm bells, so to speak, so that the government can adjust its energy policy accordingly.

In 2004, the Philippines bought 126 million barrels of oil worth US$ 4.57 billion, or about US$ 36.27 per barrel. Government estimates, however, project that every US$ 10-price increase per barrel will require an additional US$ 1.26 billion to finance our annual oil imports.[114] Imagine, with the current oil price of US$ 65 per barrel, purchasing 126 million barrels (the country's total oil imports in 2004) will cost us almost US$ 8.2 billion! As I remember it, in the past we spent 20 percent of our dollar income on oil; I do not know the ratio today, but I presume it is much higher now. At the rate the price of oil keeps increasing, where do we get the money to pay for this? We may have to export more of our labor in order to earn the much-needed dollars to finance our oil purchase.

Compared to my time, the magnitude of world energy requirements has grown tremendously, and even the economics of each energy source has changed. Before, harnessing the sun's power was very expensive. How much does it cost today? If we don't have the answer right now, then let's study it! Assuming solar power is still expensive, will it remain more expensive even when oil reaches US$ 100 per barrel? Remember, in 1972, a year before the oil shock, crude cost US$ 1.30 a barrel; in 2005, it breached the US$ 70-mark. It is not impossible that oil will reach US$ 100 per barrel since it is a finite source. How will this affect the comparative costs of other energy sources?

Generally speaking, the tapping and utilization of alternative energy is really a fruition of materials technology. Harnessing solar power is a case in point. Materials that can withstand the transformation of solar heat may now be available. If so, then there is greater potential for developing commercially viable technologies using the sun's power. To me, the utilization of solar energy is similar to that of the jet engine, which took time to develop even though it had been known as far back as a century ago that jet propulsion was the way to go. This eventually led to the creation of the most modern means of transportation to date.

Given the intimate relation between alternative energy and technological development, we should learn about what other countries are doing in developing feasible oil substitutes and determine which can be adapted to local needs. Biofuels are a good example. Biodiesel and ethanol (also known as ethyl or grain alcohol) have become viable in the

---

114. "Palace mulls fuel rationing," *The Philippine Star*, August 18, 2005.

face of escalating oil prices. Brazil, which pioneered the extraction of alcohol from sugarcane, has 320 ethanol plants that provide about 20 million transport units with 25-percent alcohol blend in their fuel tanks. With Brazil's ethanol costing the equivalent of US$ 25 per barrel of oil, Japan signed up to buy 15 million liters "as a prelude to replacing 3 percent of Japan's gasoline," which would generate an annual demand for 1.8 billion liters of alcohol. China is reportedly interested in importing Brazilian ethanol, besides building what is touted to be the biggest ethanol factory in the world. In the US, ethanol from corn has already replaced three percent of total transport fuel, while Germany has become the world's largest producer of biodiesel (made from rapeseed); what's more, major companies like Volkswagen and Du Pont are supporting large-scale research in biofuels.[115]

In my time, when PNOC experimented with alcogas, our production cost was the equivalent of US$ 30 per barrel when world oil prices were just about US$ 20 per barrel. Recently, however, our former alcogas project director Francis Lorilla (who was assigned in Bacolod, Negros Occidental) told me that with premium gasoline currently selling at almost Php 40 per liter, alcogas has become more than viable. Apparently, after the Aquino government dismantled our alcogas program, Francis continued to maintain an academic interest in alcogas and became a consultant for the sugar mills that PNOC worked with. He informed me sometime in August 2005 that, between surging oil prices and improvements in production techniques, local alcogas production has become a viable enterprise.

Unfortunately, the national alcogas program became a casualty of the Aquino government's abolition of the Ministry of Energy. Our program was intended at gradually building the country's capability for alcogas production so that we could displace 15 percent of our gasoline requirements in ten years. Had Mrs. Aquino not governed with vindictiveness and instead allowed the program to develop, we would have significantly reduced the transport sector's dependence on oil. It was another lost opportunity for the Philippines. The present government says

---

115. "The Next Petroleum," *Newsweek International,* August 8, 2005.

that we will be importing 25 million liters of alcohol for 2005, and it is urging Congress to pass a law that mandates the use of ten percent alcohol blend by 2010.[116] Why did we have to reinvent the wheel?

The cost of energy has again become a major issue for the world, and our government cannot afford to act as if it is business as usual, or parade a host of recycled solutions (like energy conservation or fuel rationing) without the benefit of proper study and evaluation. A lot of hard thinking and planning have to be made, and not just for present needs. We have to anticipate the growth in population. Assuming we reach 100 million, even at just two barrels per head, you are talking of about 200 million barrels of oil per year. Where do we get the money to pay for oil at US$ 64, US$ 80, or even US$ 100 per barrel?

In her State of the Nation Address in July 2005, President Gloria Macapagal-Arroyo asked Congress to pass a law promoting the use of indigenous energy resources. I don't mean to be rude but, excuse me, you cannot legislate energy. What has government done all these years in terms of creating the infrastructure for developing indigenous energy resources? And I really mean "creating," because the infrastructure that my team and I had built was dismantled by Mrs. Aquino, with the remainder being sold by her successors in the name of privatization. With Petron and the National Power Corporation alone, the government has lost two strategic policy instruments. Buying back Petron or reviving NPC is not a solution. In short, if we want a sound energy policy, we have to begin from scratch *again*, thanks to Cory Aquino and Fidel Ramos.

Today, any significant move in the direction of energy self-reliance will have to operate under a host of constraints that were not present in my time. Besides the fact that we are now living under a deregulated and liberalized economic regime, our heavy debt burden has restricted our room for maneuver in sourcing the much-needed capital for the energy sector. Equally important, the kind of democratic political system that we have now poses obstacles in planning, decision making, and action.

---

116. "Congress expected to pass ethanol measure next week," *The Philippine Star*, August 19, 2005.

Whether we admit it or not, the centralization of decision making under the Marcos administration turned out to be conducive for building the energy infrastructure as quickly as possible. We could not have set up all those PNOC subsidiaries in quick succession, or made important investment decisions in a matter of days, if it were not for the fact that we answered only to the president. With the democratization of politics, however, we also democratized the playing field for various interest groups that want to have their say—and share—in almost all sectors of the economy. A few of my former colleagues who have continued to work in government tell me that they were able to accomplish so much more, and in a much shorter time, during our stint at PNOC and the ministry. Very significant, too, is their assessment that we were much more professional in our conduct compared to what they experience now with the executive and legislature.

I am not saying that only a martial law regime can make possible the management of our energy sector. Neither martial law nor democracy possesses unique characteristics that will automatically make things work. Fortunately, the combination of martial law, a highly professional staff that I inherited from the old Esso (Philippines), and the confidence bestowed on me by President Marcos, enabled me to exercise my leadership in meeting the challenge posed by the oil crisis and the need to develop energy self-reliance. As Mrs. Aquino's former executive secretary and now senator Joker Arroyo told Sylvia Mayuga in 1991, "One of the good things Marcos did was putting up PNOC in the first oil crisis. We in the opposition were happy about the crisis because it created a big problem for government but Velasco did a good job."

The conditions under which I operated no longer obtain. But whatever the present constraints, it is incumbent upon government—whether the present one or the next—to exercise creativity, foresight, vision, and political will in steering our country toward energy self-reliance. There is no other way to do it, if our country is to survive rising energy costs. How do we go about it?

We have two basic uses for energy: transport and electricity generation. As far as transport is concerned, unless we discover liquid fuel or unless subsequent inventions are made that utilize other types of fuel, we will have to depend on oil. Yes, there are new developments, introduced primarily by the Japanese, regarding the use of hydrogen and rechargeable batteries for automobiles; some experimental models are already being marketed in advanced countries. But even these electric or hybrid cars are still oil-based, except that they use less oil for fuel. At any rate, while I welcome these developments in automobile engineering, I'm afraid we will still be dependent on oil for our transport needs because the Philippines has neither the expertise nor the resources to engage in similar experiments.

Now, in the case of electricity, we can maximize domestically available domestic energy such as hydro, geothermal, perhaps some alcohol and wind. The Department of Energy reports that our present energy mix is dominated by power plants run on coal (26 percent) and oil (24 percent), with the Malampaya natural gas contributing 18 percent. Wind power is beginning to attract investors and financing institutions, which is fortunate because the Philippines has 1,038 wind sites with a total potential installed capacity of 7,404 megawatts, according to the wind map survey undertaken by the US-based National Renewal Energy Laboratory. Hopefully, the US$ 400-million North Luzon Wind Power Project in Ilocos Norte, which is expected to initially generate 42 megawatts, will encourage similar endeavors in other parts of the country.

Doubtless, for the present, the cost of generating electricity via wind energy is still higher as compared to the costs of generated power via today's conventional means. In this regard, we can learn from the experience of the United States, which has a total installed wind-energy capacity of 6,740 megawatts. To accelerate the development of wind energy, interested power generators took the imaginative and innovative approach of seeking out the support of well-known American companies that publicly and consistently express concern for environmental issues. In particular, given that wind-generated electricity would cost several cents more per kilowatt-hour than, say, a coal- or oil-fired power plant, these environmentally conscious companies were asked to defray the differential of generating costs. For instance, if the companies' current electricity purchases from their grid connection cost US$ 0.12 per kilowatt-hour, and the cost of wind-generated electricity is US$ 0.15 per kilowatt-hour, the concerned companies agreed to pay the differential of US$ 0.03 per kilowatt-hour, multiplied by their total electricity consumption for the year as purchased from their current grid sources.

Such an approach accelerated the installation of wind energy throughout the US, as the communities that were to be served by wind-generated electricity did not have to shoulder the added cost, thereby eliminating consumer resistance to environmentally friendly but more expensive electricity. I do believe that this method of eliciting private sector support could be applied to the Philippines should there be a need for assistance in defraying the cost of applying new, nonconventional energy sources in the country.

The bottom line is, government has to take the lead in developing our domestic energy sources, not so much by doing the work itself but by creating the appropriate policy and business environment that would encourage private investors to go into power generation using our domestic sources. Given the high installation costs for geothermal plants or hydro systems, for instance, the government will have to provide significant tax incentives to individual power producers that will tap these sources. Our political leaders will have to be proactive in redirecting investor interest toward our non-oil sources especially because, like it or not, oil is a finite source and it will become more scarce and therefore more expensive.

The problems we confront today—high energy prices and looming power shortages in the next few years—I attribute to what I believe to be the three biggest mistakes of President Marcos's successors: the mothballing of the Bataan nuclear power plant, the sale of Petron, and the breakup of Napocor. For me, these mistakes were rooted in lack of understanding and appreciation for energy issues, lack of foresight and, most important of all, provincialism in politics.

Whatever his "sins" were, the late President Marcos could lay claim to energy development as a signal achievement. Unfortunately, Mrs. Aquino could not go beyond the vilification of her predecessor: besides discrediting the Ministry of Energy, she even dismantled the country's energy infrastructure that took over a decade to build. Other presidents have not done any better, undoing past achievements and prioritizing political careers in making decisions. Look at Napocor: people forget that the organization's bankruptcy owed much to the policy of successive post-Marcos administrations of keeping down electricity rates even if generating costs have increased.

From what I have observed since my departure from office, I can say that all post-Marcos presidents have tended to address energy issues in a piece-meal and rather uncreative manner. Let us take as an example rising oil prices.

The present government's response has been to revive alcogas, not realizing that to displace 15 percent of gasoline consumption will require ten years of developing approximately 250,000 hectares of new sugar cane fields, and the installation of new sugar centrals that will process the cane directly into alcohol. The government is also promoting coco-diesel as an alternative, without bothering to review the lessons from past experience. To me, coco-diesel is an unsuitable substitute, as the automotive engines will have to be redesigned to accommodate its use. Others suggest the repeal of the oil deregulation law or the reinstitution of the Oil Price Stabilization Fund. But how can government do either and still be effective, when it has shed the policy instruments, powers, and capabilities necessary for the competent regulation of the oil industry?

Unfortunately for the country, many of our leaders tend to resort to tired formulas without trying to learn the lessons of the past and understand the conditions of the present—both its potentials and limitations. Worse, they stubbornly hold on to the attitude of discrediting political rivals in order to enhance their own position: What to do with a successful program implemented by your predecessor in office? Look for loopholes, throw them out, and create your own. This habit of wanting to get the credit, of undoing a program that works simply because it originated from a rival or even a political enemy, cannot continue. Sadly, this is the story of the country's energy sector.

As it is now, the country has no energy policy. We have no direction, and as a consequence, we have become vulnerable to the whims of the world oil market. We cannot turn our back on energy. It is so critical to our development, and in this globalizing world, we have to exercise our collective imagination in order to cope and compete. There is no easy way out.

THE AUTHOR

GERONIMO ZAMORA VELASCO

Geronimo Zamora Velasco presided over the organization and development of the Philippine energy sector as chairman and president of the Philippine National Oil Company from 1973 to 1986, and minister of energy from 1977 to 1986. In recognition of his outstanding performance in managing the oil crisis and assuming responsibility for the energy sector, the Management Association of the Philippines declared him "Management Man of the Year" in 1977. For promoting the advancement and application of engineering, science, and technology in the field of energy, Velasco was awarded honorary doctoral degrees from the Mapua Institute of Technology, De La Salle University, Philippine Women's University, and the University of the Philippines.

Before joining government, Velasco was one of the country's foremost corporate executives, having been the chairman and chief executive officer of Republic Glass Corporation (1959-1973), Dole (Philippines) Inc. (1965-1973), and Gervel Inc. (1966-1973). He also served as director of several Philippine and foreign companies such as Philippine Rock Products Inc., General Motors (Philippines) Inc., RCA (Philippines) Inc., Manila Memorial Park, Castle & Cooke Investment Company (Honolulu), Granite Industries Berhad (Malaysia), and Oceanic Properties Investments Inc. (Singapore), among others. He is currently the chairman of Republic Glass Holdings Corporation and Gervel Inc.